SpringerBriefs in Environment, Security, Development and Peace

Volume 32

Series Editor

Hans Günter Brauch, Peace Research & European Security Studies, Mosbach, Baden-Württemberg, Germany

More information about this series at http://www.springer.com/series/10357
http://www.afes-press-books.de/html/SpringerBriefs_ESDP.htm
http://www.afes-press-books.de/html/SpringerBriefs_ESDP_32.htm

Yoshitsugu Hayashi · Masafumi Morisugi ·
Sho-ichi Iwamatsu
Editors

Balancing Nature and Civilization—Alternative Sustainability Perspectives from Philosophy to Practice

 Springer

Editors
Yoshitsugu Hayashi
Center for Sustainable Development
and Global Smart City
Chubu University
Kasugai, Japan

Masafumi Morisugi
Faculty of Urban Science
Meijo University
Nagoya, Japan

Sho-ichi Iwamatsu
Graduate School of Environmental Studies
Nagoya University
Nagoya, Japan

Translated by SIA Inc., Nagoya, Japan, Junichi Kimura, Hiroshi Miyata, Asako Shibagaki, Sumie Tsuge, Paul Mason and Kenji Sasaki

More on this book is at: http://www.afes-press-books.de/html/SpringerBriefs_ESDP_32.htm

ISSN 2193-3162   ISSN 2193-3170   (electronic)
SpringerBriefs in Environment, Security, Development and Peace
ISBN 978-3-030-39058-7   ISBN 978-3-030-39059-4   (eBook)
https://doi.org/10.1007/978-3-030-39059-4

Copyediting: PD Dr. Hans Günter Brauch, AFES-PRESS e.V., Mosbach, Germany

This Springer imprint is published by the registered company Springer Nature Switzerland AG
The registered company address is: Gewerbestrasse 11, 6330 Cham, Switzerland

Preface

This book is a concise record of an international symposium: *Sustainability—Can We Design the Future of Human Life and the Environment?* The symposium brought together world-renowned researchers and was hosted by the Graduate School of Environmental Studies, Nagoya University. The symposium was held as a satellite event of the "*Love the Earth*"-Expo 2005 (Aichi, Japan) on 6 August 2005. August 6 is the day the atom bomb was dropped on Hiroshima, and two authors (Ishii and Dürr) remark on it in their introductions.

Chapter 1 is the introductory section. Dr. Yoshitsugu Hayashi, the chairperson of the symposium, explains the changes in the global environment and how we should keep an environmental balance between nature and human activities.

Part I of this book, titled *A Sustainable Relationship Between Nature and Humans*, is a collection of presentations delivered by four scholars. The presenter of Chap. 2 is Dr. Yoshinori Ishii. He is a geophysicist, a resource scientist and the founder of the Mottainai Society in Japan. He explains the finite nature of natural resources, peak/depletion of oil, the importance of the energy profit ratio (EPR), and post-oil strategies such as local production for local consumption.

The presenter of Chap. 3 is the late Dr. Hans-Peter Dürr. He was a successor to Dr. Werner Heisenberg in nuclear physics. He was also a philosopher who compiled the Potsdam Manifesto 2005, a follow-up to the Russell–Einstein Manifesto of 1955. He warns that since the industrial revolution our human race has been behaving like a bank robber towards our planet Earth by excavating fossil resources beyond a sustainable amount, until Earth has lost its original eco-balance. He explains how the Earth has stored resources with the benefit of solar energy, and how we should use energy to keep the dynamic Earth's ecosystem sustainable.

The presenter of Chap. 4 is Dr. Yoshinori Yasuda. He is an environmental archaeologist and a geologist. Based on his pollen analyses, he explains how food cultures impact on forests and ecosystems, showing a stark contrast between European and "monsoon Asia" civilizations, and how we should avoid a future collapse of modern civilization caused by overpopulation and the Western culture in which human diet is dominated by meat, which leads to domesticated animals eating higher orders of plants than we eat directly.

The presenter of Chap. 5 is Dr. Minoru Kawada. He is a politics scholar and a historian. Based on the discussions of the Japanese view of ethics and faith by Kunio Yanagita, the father of Japanese folklore study, he explains how *Ujigami* (local small god) worship and intergenerational ethics have preserved Japanese forests and nature. In traditional villages in Japan, three generations lived together in a family. There was an oral tradition of grandparents telling their grandchildren that a god lived in the mountain behind the village and that she was so great that no human could challenge her.

Part II of this book, titled *International Conflict Concerning Environmental Damage and Its Causes*, collects presentations delivered by two scholars. The presenter of Chap. 6 is Dr. Yasunobu Iwasaka. He is an atmospheric physicist and climate scientist. He explains the history of international research on *Kosa* (Asian dust particles) which includes *Kosa* as a tracer of air pollutions, evaluations of the effect of *Kosa* on global warming, and other matters. He then discusses what kinds of conflict and cooperation have taken place among Asian countries in this international joint research.

The presenter of Chap. 7 is Dr. Werner Rothengatter. He is a transport economist and a traffic engineer in Germany. He explains environmental charges on heavy goods vehicles (HGV) in the EU, including different toll systems between countries, emission standards, and categories. He discusses how EU countries have formed a consensus under different geographical and economic environments. The data in this chapter is updated for this publication.

Part III of this book (Chap. 8), titled *Ecological Balance and Conflicts in the 21st Century*, relates to opinions by the designated speakers and a panel discussion chaired by Dr. Hisae Nakanishi. Three designated speakers (Dr. Dongyuan Yang, the late Dr. Lee Schipper, and Dr. Itsuo Kodama) express their opinions based on viewpoints of environmental conflicts inside China, tax policy to prevent energy wastage, and the relationship between the environment and human health, respectively. Following the above opinions, presenters' answers to questions from the floor are summarized in the Questions and Answers section.

Each chapter is easy to read, with detailed and clear descriptions. I hope that the book's great insights will come across to readers, and be of help in steering the world towards a more sustainable society in harmony with the Earth's biosystems.

I would like to thank the committee members of the Graduate School of Environmental Studies who organized the symposium. I also express my appreciation to the co-supporters of the symposium: Academic Consortium 21 (AC21), UFJ Environment Foundation, the *Chunichi Shimbun* Newspaper, Aichi Prefecture, Aichi Prefectural Board of Education, Nagoya City, Nagoya City Board of Education, and NHK Nagoya Broadcasting Center.

This book is basically an English translation of the Japanese version of 2010, and I must not forget to mention Mr. Paul Mason and Mr. Kenji Sasaki of SIA Co., Ltd. for their excellent translation. I would like to extend my gratitude to each of these individuals and organizations.

Nagoya, Japan   Yoshitsugu Hayashi
May 2019   Chairperson of the Executive Committee
                                                                                                    of the Symposium
                                                                                   and Chief Editor of this Book

Contents

1 Introduction: Can We Design the Future of Human Life
and the Environment? 1
Yoshitsugu Hayashi

Part I   A Sustainable Relationship Between Nature and Humans

2 The Fate of Twentieth-Century Civilization – A Discussion
of "Post-oil Strategies" 9
Yoshinori Ishii

3 Sustainable Use of Energy 19
Hans-Peter Dürr

4 Sustainability from the Perspective of Environmental
Archaeology 33
Yoshinori Yasuda

5 Re-evaluating the Traditional Japanese Perspective
on Nature and Ethics 51
Minoru Kawada

**Part II   International Conflict Concerning Environmental
Damage and Its Causes**

6 *Kosa* (Asian Dust Particles) and Air Pollution in Asia 65
Yasunobu Iwasaka

7 Environmental Charges Levied on Heavy Goods Vehicles
in the EU 77
Werner Rothengatter

Part III   Ecological Balance and Conflicts in the 21st Century

8   Panel Discussion   95
      Chair: Hisae Nakanishi; Designated speakers: Yang Dongyuan,
      Lee Schipper, Itsuo Kodama; Panelists: Yoshinori Ishii,
      Hans-Peter Dürr, Yoshinori Yasuda, Minoru Kawada,
      Yasunobu Iwasaka and Werner Rothengatter

About the Editors   111

About the Authors   113

About the Book   119

List of Figures

Fig. 1.1 Earth: compared to a cliff top. *Source* The author 1
Fig. 1.2 'Sustainability' in the Graduate School of Environmental Studies. *Source* The author 2
Fig. 1.3 Visualization of mankind's resource usage. *Source* The author 3
Fig. 1.4 Visions and processes of sustainability studies. *Source* The author 4
Fig. 2.1 Limited earth. *Source* Y. Ishii, June 27, 1984. © The author 10
Fig. 2.2 Trends in annual production of all hydrocarbons with projections to 2050. Gboe: Gigabarrels of oil equivalent. *Source* Campbell (2002). Reprinted with permission from Ecotopia 13
Fig. 2.3 Oil – past discovery/production & projected depletion; CO_2 emissions–past consumption and projected control (ASPO, GCI). *Source* The Global Commons Institute, GCI (2003). Reprinted with permission from Contraction & Convergence (C&C) ©®™ 14
Fig. 3.1 Geo-ecosystem. *Source* The author 20
Fig. 3.2 Solar energy. *Source* The author 21
Fig. 3.3 Sun as source of syntropy. *Source* The author 22
Fig. 3.4 Islands of high syntropy on earth. *Source* The author 22
Fig. 3.5 Bank robber. *Source* The author 23
Fig. 3.6 Biosystem. *Source* Photo editing by Seidel and Weidlich 24
Fig. 3.7 Dynamic meaning of sustainability. *Source* The author 24
Fig. 3.8 World economy. *Source* The author. Inspired by Herman Daly (1990) 26
Fig. 3.9 Key role of energy. *Source* The author 27
Fig. 3.10 Personal energy consumption (USA). *Source* The author 28
Fig. 3.11 Personal energy consumption (Europe, Japan, China, Bangladesh). *Source* The author 29

Fig. 3.12   Carbon dioxide production and energy slaves per person.
             Source The author   29
Fig. 4.1   Fossil palm pollen from Easter Island. *Source* The author   34
Fig. 4.2   Climate division of Eurasia. *Source* Yasuda (2002: 51).
             Reprinted/adapted with permission from Chuokoron-Shinsha,
             Inc. © 2002 Chuokoron-Shinsha, Inc./Yoshinori Yasuda   34
Fig. 4.3   Relative pollen diagram from the Ghab Valley, Syria.
             Source Yasuda et al. (2000: 131). Reprinted/adapted with
             permission from Elsevier Science Ltd. and INQUA.
             © 2000 Elsevier Science Ltd. and INQUA   36
Fig. 4.4   Bald mountain around Ghab Valley. Forests and water
             completely disappeared from the West Asian Fertile Crescent.
             Source The author   36
Fig. 4.5   Selected pollen diagram on the top 16.2 m part of a 120 m core
             from Lake Copais, Southeast Greece. *Source* Okuda et al.
             (1997: 109). © The Author(s).   37
Fig. 4.6   Greek islands and the blue Aegean Sea. Destruction of forest
             stopped flow of nutriments so sea became barren.
             Source The author   38
Fig. 4.7   Climate change restored from the $\delta^{13}C$ record of a giant cedar
             tree from Yakushima Island, southern Japan. *Source* Kitagawa
             (1995: 50). Reprinted/adapted with permission from Asakura
             Publishing Co., Ltd. © 1995.   40
Fig. 4.8   Distribution of forest in USA. **a** In 1620, **b** In 1920.
             Source Goudie (1993: 45). Reprinted with permission from
             Blackwell Publishers. © Goudie 1981, 1986, 1990, 1993.
             See also Yasuda (2002: 7)   40
Fig. 4.9   Yuchanyan in Hunan Province, China. Rice cultivation began
             around 14,000 years BP. *Source* The author   41
Fig. 4.10   Terrace paddy field in Guizhou Province, China. Rice farmers
             devote their energy to creating fertile land.
             Source The author   42
Fig. 4.11   Moai statue on Easter Island (El Gigante).
             Source The author   45
Fig. 4.12   Mother goddess in Tahiti. *Source* The author   45
Fig. 4.13   The forests of New Zealand were destroyed after the arrival
             of Europeans. Left: Forest distribution in the 7th century AD.
             Right: Forest distribution in 1974. *Source* Goudie (1993: 33).
             Reprinted/adapted with permission from Blackwell Publishers.
             © Goudie 1981, 1986, 1990, 1993   46
Fig. 4.14   Collapse of present civilization. *Source* Meadows et al.
             (1972: 124) CC BY-NC; Adapted and rewritten by Yasuda
             (2005: 89). Reprinted/readapted with permission from Wedge
             Inc. © Yasuda, Kobayama, Matsui   47

Fig. 4.15   Futamigaura in Ise City, Mie, Japan. The sun rises between the
             Married Couple Rocks. The larger rock has a small *torii* at
             its peak. *Source* The author   48
Fig. 5.1   Kunio Yanagita. *Source* Kawada (1997)...................   51
Fig. 5.2   *Ujigami* worship and intergenerational ethics (festival of
             ujigami). *Source* Photo courtesy of Koichi Kato.............   56
Fig. 6.1   History of *kosa* research in Japan. *Source* The author   66
Fig. 6.2   *Kosa* observed with Lidar at Nagoya (April 1979).
             Source Iwasaka et al. (1983: 191) CC BY 4.0   68
Fig. 6.3   Air masses from China observed with Lidar at Nagoya
             (April 1979). Curve A was computed for the potential
             temperature $\theta = 316.4$ K and curve B for $\theta = 292.2$ K.
             Source Iwasaka et al. (1983: 193) CC BY 4.0   69
Fig. 6.4   The IPCC third report. The global mean radiative forcing
             of the climate system for the year 2000, relative to 1750.
             The term "Mineral Dust" appears in the middle as a component
             of Aerosols. *Source* IPCC (2001: 8)......................   69
Fig. 6.5   Concentration of atmospheric CO_2 measured at Mauna Loa,
             Ryori and South Pole. *Source* The Japan Meteorological
             Agency (2004: 31)   70
Fig. 6.6   Change in standing point of *kosa. Source* The author.........   71
Fig. 6.7   Balloon-borne observation and ground-based Lidar
             observation at Dunhuang with Japan/China cooperation.
             Source The author   72
Fig. 6.8   Balloon-borne particle sampler. *Source* The author...........   73
Fig. 6.9   Particle sampler mounted on balloon. *Source* The author......   74
Fig. 6.10   Relative weight ratios of Al, S and Ca for the mineral particles
             collected in Dunhuang (**a–e**) and in the free troposphere over
             Japan (**f**). *Source* Trochkine et al. (2003: 8). Reproduced and
             modified by permission of American Geophysical Union.
             © American Geophysical Union, 2003   75
Fig. 7.1   Variety of HGV charging systems in the EU.
             Source EU Commission (2015)   80
Fig. 7.2   Toll collect on-board unit for a 5-axle truck in Germany.
             Source Toll Collect (2019a)   84
Fig. 7.3   Cost allocation scheme for German motorways.
             Source Rommerskirchen et al. (2009). © The Author(s).......   85
Fig. 7.4   European Electronic Toll Service (EETS).
             Source European Commission (2019)......................   89

List of Tables

Table 7.1   Charges for HGVs in Switzerland since 2017   81
Table 7.2   HGV tolls in Austria for 2019   83
Table 7.3   Charges for infrastructure costs in Germany since 1/2019   86
Table 7.4   Charge for environmental costs in Germany since 1/2019   87
Table 7.5   EURO emission standards for HGV   88

Chapter 1
Introduction: Can We Design the Future of Human Life and the Environment?

Yoshitsugu Hayashi

What has become of the global environment? A very important factor in the development of people's awareness of the global environment was the emergence in the later 1960s of the concept of Spaceship Earth. Issues of air or water pollution came to people's attention in many places on Earth. Japan also experienced serious environmental pollution in the 1970s. People began to consider that Earth's capacity to purify pollution was not infinite, but had limits.

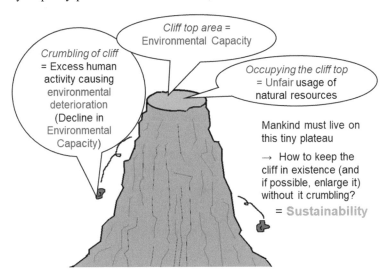

Fig. 1.1 Earth: compared to a cliff top. *Source* The author

Yoshitsugu Hayashi, Chair of the Symposium; Professor and Director, Center for Sustainable Development and Global Smart Cities, Chubu University, Japan; Professor Emeritus and former Dean, Graduate School of Environmental Studies, Nagoya University, Japan. Email: y-hayashi@isc.chubu.ac.jp.

However, in my opinion, rather than simply considering that Earth's purifying capacity is not finite, we should visualize ourselves as living in a limited area on the top of a cliff, as illustrated in Fig. 1.1. In addition, the area not only has the limits suggested by Spaceship Earth: rather its edges are gradually crumbling. In other words, the area available for use is decreasing as a result of pollution, etc. What will happen next? Those furthest from the center of the limited area, who are most vulnerable, fall off the cliff. It may not be humans, but other creatures that fall off. In my opinion, it is now necessary to change our view and think about how to maintain the area on the top of the cliff and prevent the cliff from giving way.

These ideas induced us to open the Graduate School of Environmental Studies (at Nagoya University), and Fig. 1.2 illustrates what we came up with then. It illustrates the human race, walking a tightrope, trying to keep a balance between 'Nature' and 'product', which represent respectively things produced by Nature and humanity. We have chosen this illustration as a model of sustainability because it depicts humans walking a tightrope that stretches into the future, trying to balance Nature and products well to ensure safety and security. How to theorize safety, security and sustainability is an important research topic on which our Graduate School of Environmental Studies works.

If we graphically represent how humans use resources (Fig. 1.3) as an example of sustainability, it illustrates how the abundance of resources on Earth gradually decreases, or how the cliff gives way. If we continue using the depleting resources with current technologies, under current rules or with current awareness, our standard of living could reach its peak, but drop sharply afterward, as shown by the dotted trend curve in the figure. It is also possible that our standard of living is about to pass its peak. We need to think: how do we prevent the cliff from rapidly giving way in the future; how do we live a life like a human being; and how do we

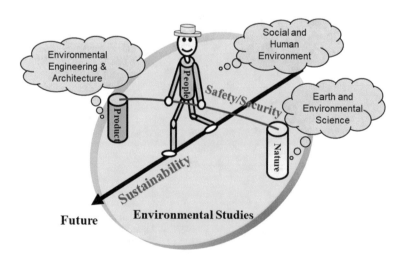

Fig. 1.2 'Sustainability' in the Graduate School of Environmental Studies. *Source* The author

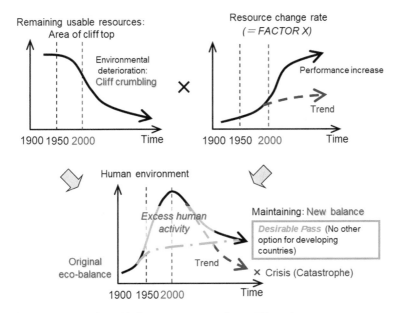

Fig. 1.3 Visualization of mankind's resource usage. *Source* The author

co-exist with non-human, terrestrial life. There is a key to approaching these issues: the energy conversion efficiency we call Factor X. If the amount of human activities doubles but the efficiency of resource use fails to double, twice the resources will be consumed, and twice the damage will be done to Earth's environment.

We are trying to chart this issue in terms of the visions and processes of sustainability studies (Fig. 1.4) to understand it better. It is often said that human activity and the natural environment may have been in a primitive environmental equilibrium in pre-industrial agrarian society. However, excessive human activities cause an imbalance, and further development of the imbalance produces conflicts and crises. In such a case, we have to look for a way to keep the balance. The restoration of the original, primitive equilibrium is no longer feasible because we would have to go back to the lifestyle of primitive times to restore that equilibrium.

To look for a new way of achieving an environmental balance, it is necessary to consider what indicators should be used to monitor changes and the collapse of the balance. In addition, when considering this issue, values held by people, such as their views on Nature and ethics, play a very important role.

In order to monitor the changes, we must set goals and benchmarks as interim targets. Then, we have to look for alternative soft-landing paths, or strategies, to reach the goals, as well as sets of necessary technologies and policies. Ultimately, to implement the policies and shift to a new balance, it is necessary to achieve a consensus not only in Japan, but also on a global basis in developed and developing countries.

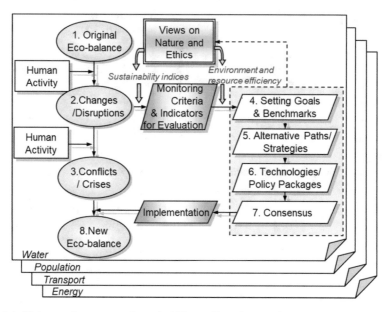

Fig. 1.4 Visions and processes of sustainability studies. *Source* The author

We must keep an eye on the balance between Nature and civilization, and control it. There are different factors, such as water, food, population, traffic and transportation, and energy, to discuss how to do it, and it should be discussed in terms of each. Although it is difficult to understand these issues correctly, I think it important to categorize them before attempting to understand them.

References

Hayashi Y, Yasunari T, Kanzawa H, Kato H (eds.) (2016) *Climate Change, Energy Use, and Sustainability: Diagnosis and Prescription after the Great East Japan Earthquake.* Springer, Switzerland

Matsuo M, Hayashi Y (2001) Need for Changing Converters of Civilization for 21st Century (in Japanese). *Japanese Scientific Monthly* 6:615–619

von Weizsaecker EU, Hargroves C, Smith MH, Desha C, Stasinopoulos P (2009) *Factor Five, Transforming the Global Economy through 80% Improvements in Resource Productivity.* Earthscan, London

von Weizsaecker EU, Wijkman, A (2018) *Come On! Capitalism, Short-termism, Population and the Destruction of the Planet – A Report to the Club of Rome.* Springer, New York

Yoshitsugu Hayashi (Japan), born in 1951, Professor Emeritus, Nagoya University; and Director, International Research Center for Sustainable Development and Global Smart Cities, Chubu University; Distinguished Visiting Professor, Tsinghua University, China. He is a Full Member of the Club of Rome and President of the Japan Chapter, and also has been President of WCTRS (World Conference on Transport Research Society) till May 2019.

His major fields of research are analysis and modelling of transport – land use interactions and the countermeasure policy to overcome negative impacts of urbanization and motorization. The results are published in such books as *Land Use, Transport and The Environment* (Kluwer, 1996), *Urban Transport and the Environment – An International Perspective* (Elsevier, 2004), *Intercity Transport and Climate Change – Strategies for Reducing the Carbon Footprint* (Springer, 2014), the Japanese Edition of *Factor 5* (Akashi-shoten, 2014) originally authored by Ernst Ulrich von Weizsaecker, et al., section author of *Come on: Capitalism, Short-termism, Population and the Destruction of the Planet* (Club of Rome Report, Springer, 2018) edited by Weizsaecker and Wijkman.

Applications to practice include his proposition of rail transit-oriented urban reform to overcome Bangkok's hyper congestion as the leader of JICA project in the mid-90s, which became the trigger to reverse the budget of road versus rail from 1:99 in the 90s to 82:14 in the Transport 2020 Plan. He is also now JICA/JST research project leader of "Smart Transport Strategy for THAILAND 4.0".

Part I
A Sustainable Relationship
Between Nature and Humans

Chapter 2
The Fate of Twentieth-Century Civilization – A Discussion of "Post-oil Strategies"

Yoshinori Ishii

On August 6, sixty years ago, an atomic bomb was dropped on Hiroshima. I was then 12, living in Urawa City, Saitama. I still vividly remember the newspaper article reporting that a new bomb had been dropped. How could America have atom-bombed Hiroshima, a city where hundreds of thousands of people were living? Even as a child, I thought it was a criminal act. It was said that Japan would not have surrendered without the atomic bombing. So "Why did they not drop it in the sea: in the Seto Inland Sea for example?" I thought as a child. Human beings still have a large number of nuclear weapons. Men do not seem to have got any wiser.

My concern is what will become of 20th-century-type civilization. 20th-century civilization was sustained by oil. But depletion of oil supplies is now in sight. In April this year (2005), Prime Minister Koizumi used the term "post-oil strategy," which later appeared on the Cabinet website. This was the first time that the top leadership of a country had declared that the supply of oil was limited, and that the Earth itself is limited. I therefore chose the subtitle "A Discussion of Post-Oil Strategies" for this lecture. I would like to discuss possible sustainable ways of living for the rest of the 21st-century.

I am afraid it is difficult to express my thoughts within today's limited time so they can be fully understood. So please take a look at my personal website (see the author profile at the end of this chapter).

2.1 Limited Earth

Look at the first figure (Fig. 2.1). I drew this 20 years ago, in 1984, when I was a professor of the University of Tokyo. The jagged lines indicate that it was drawn 20 years ago. I remember I was the first to buy a Macintosh at the University of

Yoshinori Ishii, President, Mottainai Society (NPO); Professor Emeritus, University of Tokyo, Japan; former Professor of Toyama University of International Studies, Japan.

© The Author(s), under exclusive licence to Springer Nature Switzerland AG 2020
Y. Hayashi et al. (eds.), *Balancing Nature and Civilization—Alternative Sustainability Perspectives from Philosophy to Practice*, SpringerBriefs in Environment, Security, Development and Peace 32, https://doi.org/10.1007/978-3-030-39059-4_2

Fig. 2.1 Limited earth.
Source Y. Ishii, June 27,
1984. © The author

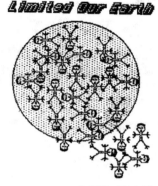

(Y. ISHII, 1984.6.27)

Tokyo. People are spilling out of the globe. This demonstrates that resources, energy, the environment, etc. are limited.

Since I specialize in geophysics, I am absolutely certain that the Earth is round and limited. But the general public seems to think that the Earth is unlimited, or they think that advances in science and technology will help overcome the limits, or market principles will somehow sort it out. In 1984, the world population was only 4.4 billion, which was enough to make me realize that human activity is exceeding the capacity of the Earth. Now the population has reached 6.4 billion. Although global warming was not clearly recognized in those days, I have spent the last 20 years continually teaching that the Earth is limited.

Now that the limits of the global environment and natural resources have become global-scale issues, humans have become no wiser. With the population reaching 6.4 billion, the situation is getting even more serious.

2.2 Eclipse of Petroleum Civilization

The main point of my talk is that oil is limited. Oil production will not keep up with demand. Today the price of oil is unbelievably high at around 60 dollars per barrel. Many Japanese people, including government leaders, however, think or want to think that this is a temporary trend. But it is not the case.

First, oil production is nearly reaching its peak. This is called "peak oil." Some mention this term as a past event, but I think peak oil will take place before 2010. Many argue that we can rely on natural gas when oil is gone. But natural gas is similar to oil, and will last until around 2020 at maximum. In this sense, we see the eclipse of the 20th-century "petroleum civilization."

Oil supports not only energy, but also agriculture and food. Fertilizers and pesticides are synthetically made from oil, while agricultural machines are powered by oil. The oil issue is the food issue at the same time. Moreover, a large amount of plastic – which is also made from oil – is used in this building around us. This means that a significant problem will arise in the basic raw materials we use today.

Thus, the first point is to see peak oil as an implication that 20th-century civilization will not prevail. In Japan, we have solely considered the quantity of oil but not paid much attention to the quality. This has led to the notion that we still have plenty of it. So, the second point is to think about the quality of resources.

The third point is to think how human beings should live in the future. There are no clear answers to this. But there are some clues to finding them, which I would like to organize in my talk. I will talk in particular from the standpoint of Japan and Asia.

2.3   Waste Is Not a Resource

Humans consume a large quantity of resources and energy for their mass production, mass consumption, and mass dumping. This results in environmental problems. In Japan, people seem to believe that recycling solves the problem. Waste is seen as a resource if it is recycled.

However, this should be considered very carefully. Massive distribution in industrial society boosts entropy. Human activity is a process that constantly increases entropy. To turn this process back, which means to lower entropy, always requires energy. Zero-emission in a strict sense may refer to a society of infinite energy. In short, complete recycling cannot be easily achieved. In this sense, energy is extremely important in modern society.

2.4   History of the Amount of Oil Discovered

Many say that there is still plenty of oil because the Earth is large. Historically, the amount of oil discovered has fluctuated sharply, as some supergiant oilfields located mainly in the Middle East constitute most of the total. On average, however, world oil discovery peaked around 1964. This means that now, in 2005, we are using an energy source that was discovered quite a long time ago. Many of the supergiant oilfields are 60 or 70 years old. Human beings are supported by past discoveries.

After reaching its peak, the amount of oil discovery has continued to decline, and the subsequent price rise is therefore structurally natural. Today the price has reached around 62 dollars per barrel. While oil use keeps rising, the amount of oil discovery now is a quarter of oil use, hardly meeting demand.

The amount of oil discovery has declined sharply; 35 gigabarrels (giga means a billion) per year between 1945 and 1960, 23 gigabarrels per year between 1970 and 1990, and 6 gigabarrels per year between 1990 and 1999. Meanwhile, annual usage between 1990 and 1999 was 25 gigabarrels, about four times the amount discovered. This situation has been continuing to date.

Many say that we can somehow manage the situation, because the Earth is big, if market principles work out, or if science and technology advance. But the history of oil discovery tells us that it is impossible. Oil reserves are concentrated in the Middle East, which is a unique area on the planet. The Ghawar Oil Field of Saudi Arabia, the largest oilfield in the world, produces 4.5 million barrels of crude oil every day. This accounts for 60% of total oil production in Saudi Arabia. This oilfield, discovered around 1948, is so old that it maintains its flowing pressure with an injection of seven million barrels of seawater each day. The crude oil produced contains around 30% water. If Ghawar, the world's largest, falls, Saudi Arabia will fall. If Saudi Arabia falls, the world will fall.

The world's second largest oilfield, the Burgan Oil Field, is located in Kuwait. This oilfield, burned during the Gulf War, was discovered in the 1930s. The Kirkuk Oilfield in Iran is even older, discovered in the 1920s. We often see this oilfield on TV. Thus, supergiant oilfields in the world are concentrated in the Middle East and they are all very old.

Other major oilfields, such as the Daqing Oilfield, the largest oilfield in China, and the North Sea oil fields, are also well known, though the scale of each field is small and together they are still smaller than a mid-scale field in the Middle East. It should be understood that the Middle East is a special region. To be specific, nearly two-thirds of world oil production, 685.6 billion barrels, is in the Middle East. Recently oilfield development has been very active in the Caspian Sea, for example. But recoverable reserves from this area are estimated to be barely 30 billion barrels. The amount of oil used by human beings in a year equals the total reserves of the entire Caspian Sea area. Human beings really use a huge amount of oil. Thus, in the end the Middle East will be the only area that remains.

2.5   Peak Oil

In 1998, C. J. Campbell, petroleum geologist, announced his forecast of world oil reserves in *Scientific American*. It was represented as a Hubbert Curve. The curve shows that the peak of world oil production comes in 2004, which attracted various objections. In fact, it does not matter whether the peak comes in 2004 or 2005. Because the curve is smooth, it is enough to know that the peak comes some time before 2010.

M. Simmons, energy adviser to US President Bush says that the peak came in 2005. I met him at this year's ASPO conference in Lisbon. ASPO stands for The Association for the Study of Peak Oil & Gas. Volunteer representatives of 14 European countries meet every year; in Berlin last year and in Paris the year before. I was the only Asian participant at the meeting in Paris.

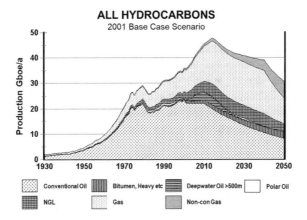

Fig. 2.2 Trends in annual production of all hydrocarbons with projections to 2050. Gboe: Gigabarrels of oil equivalent. *Source* Campbell (2002). Reprinted with permission from Ecotopia

Today, consumption of natural gas is rapidly increasing, but it is expected to reach its peak around 2015. Other energy sources, such as heavy oil, have been drawing attention, but as Fig. 2.2 shows, all hydrocarbons will be depleted.

I would like to emphasize here that "peak oil" does not mean that oil is running out, but that oil is being depleted. In this sense, we think that peak oil has already come.

ExxonMobil, the world's largest oil company, states in a document released to the public in 2004 that oil demand will keep growing while production and supply will decline, and enormous investments will be required to fill this gap. According to this document, the rate of production decline is 4–6% per year and the production peak will be before 2005. This means that ExxonMobil has declared that we have already reached peak oil.

It is interesting to compare this statement with Campbell's Hubbert Curve described earlier. Both say the same things about oil reserves and oil production. The difference is that ExxonMobil argues that it is possible to fill the gap between supply and demand, like a goal to aim at.

In Japan people seem to think that growing demand can be somehow satisfied. Our opinion is that it cannot be. Even with technological advances, it is theoretically impossible because the Earth is limited. ExxonMobil is a private company. But the International Energy Agency, an international organization, said basically the same thing as ExxonMobil in 2004.

2.6   Impact of Oil Depletion

Japan is a country where people easily believe superficial stories released in public, such as those based on official data, and they tend to develop strategies from a public standpoint. But a public standpoint often differs from the reality. If you have official data in front of you, the point is whether you have the ability to see what it really means, or whether you have an insight into the data. We should think very seriously, even working on the assumption that common knowledge is always wrong.

In considering resources and energy, EPR (Energy Profit Ratio) is important. It refers to the ratio of energy input to energy output. Money is no longer a key indicator. For more details on this, please see my website.

Another important notion is that the problem of oil is also the problem of food. Today Japan's food self-sufficiency rate is only 40%; it used to be higher. The self-sufficiency levels of the UK and Germany, which used to be low, are now quite high. What will Japan's food supply be like when peak oil comes? Our means of transportation also depend on oil. We bring food from even the other side of the earth. We import vegetables from China. What should we do if shipping charges rise even higher? Cars and airplanes are not exceptions. Thus peak oil will have its first impact on transportation.

It is important to think about how to hold down the increase in cumulative CO_2 emissions. If the amounts of cumulative consumption of oil, natural gas, coal, etc. are projected, however, we can estimate the quantity of CO_2 emissions. ASPO, which I mentioned earlier, projects lower emissions than the IPCC (International Panel on Climate Change) does (Fig. 2.3). There should be a big strategic difference between tackling global warming based on the assumption that there is plenty of oil and doing so based on the assumption that peak oil is coming soon.

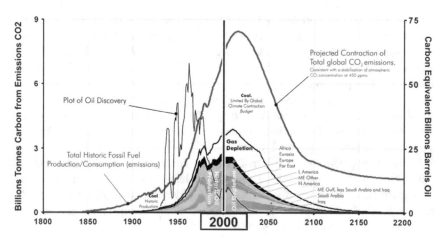

Fig. 2.3 Oil – past discovery/production & projected depletion; CO_2 emissions–past consumption and projected control (ASPO, GCI). *Source* The Global Commons Institute, GCI (2003). Reprinted with permission from Contraction & Convergence (C&C) ©®™

2.7 Resources

Please understand that a resource must be something that is concentrated, abundant, and economic. Mere abundance is no use. Natural energy sources, such as solar energy, are abundant and present in front of us but not concentrated. This is the largest reason that usage of natural energy has not progressed as rapidly as hoped. For resources, quality is everything. We must consider EPR, or Energy Profit Ratio.

Various energy sources have been proposed. Natural gas is limited. As for nuclear power, nuclear fission on a once-through basis is included in the scenario of Plan B (described below). But nuclear fusion and ITER (International Thermonuclear Experimental Reactor) are still a long way from being put into practice. It cannot be brought on line in time for peak oil. Coal was replaced by oil because oil was more practical. So coal cannot become the savior after the depletion of oil.

Oil sand, Orinoco tar, oil shale, etc. should not be considered as an extension of oilfields. Oil shale, very like mining, consists of rocks from which heavy crude oil is extracted by hot-steam injection. As it is still in the form of tar, the addition of hydrogen or light oil is necessary. Thus it is a big mistake to expect such substances to be a substitute for oil.

Natural energy sources are broadly distributed. Understanding this fact, we need to utilize them as much as possible in the future. Methane hydrate and solar power from space are unrealistic. Speaking of a hydrogen society, the question is what hydrogen will be made from. Simply suggesting hydrogen will not solve the problem.

2.8 Plan B: Post-oil Strategies

A book by R. Heinberg, *The Party's Over,* is being widely read these days. Heinberg said, in a US lecture titled Powerdown, that the world is entering an age of relocalization. It is the concept of building a society based on local food, local water, and local transportation as much as possible, as well as local economy, and local procurement of essential goods. This concept corresponds to Japanese terms such as *chiho bunsan* ('decentralization') and *chisan chisho* ("local production for local consumption"). They mean the same thing.

Do you know Plan B? Plan B refers to a transitional process that bridges Plan A, which is based on the current energy system, and Plan C, a future sustainable system. In seeking a soft landing for the age of oil depletion, time frames are very important. The term "Plan B" does not always refer to the same system. Lester Brown published a book called *Plan B*, which has a problem with its energy strategy. My Plan B differs from his perspective. Matt Simmons uses the term Plan B in the same sense I do.

While many people still do not recognize "peak oil," it should be identified as an issue of national risk management or security. We have argued repeatedly that whether you believe it or not, it is an issue of risk management. Maybe as a result of this, Prime Minister Koizumi used the term "post-oil strategy." As it is already clear to you, energy strategy must be developed based on EPR.

It is necessary to fundamentally review modern agriculture and food infrastructure. Nature-friendly *chisan chisho* (local production for local consumption) and agriculture free from fertilizers and pesticides made from fossil fuels should be promoted. In this regard, the cases of North Korea and Cuba teach us a lesson. When the supply of oil from Russia was suspended, North Korea starved while Cuba did not, because Cuba had returned to natural agriculture. Post-oil chemical raw materials are another issue we should keep in mind.

From now on, we must seek a way to coexist with nature. Oil enabled urban concentration in the 20th-century. In the 21st-century, therefore, we have to pursue "coexistence with nature" and 'decentralization.' We will also develop science and technology based on these principles, and we will communicate them as a message from Japan to Asia. Incidentally, coexisting with nature is said to be the origin of oriental thought. 20th-century urban-centered civilization may have been based originally on Western ideas. Since the Meiji era Japan has introduced these ideas as the direction for its modernization. Now we are seeing their limits.

Japan is not on a continent but it is a small archipelago, 75% of which is covered by mountains. From the geographical perspective, it is about time for Japan to change its continent-oriented direction. Transportation is an urgent issue. Airplanes, cars and ships are all powered by oil. It is important to note that oil is liquid at normal temperature. This is well-known and the most significant characteristic of oil. Because it is liquid, it can be used in internal combustion engines. But this oil is reaching its limit. In Japan, we have railways and also subways. In Tokyo, you can go anywhere without using a car. Japan's railways, including tramways, should be re-evaluated.

Finally, let me introduce a document prepared by the governor of the US state of Oregon in 1975 saying that the current age of fossil fuels would sooner or later end. In the time scale from 10,000 years ago to 10,000 years from now, human beings are currently in the age of fossil fuels, and the peak of prosperity will reach its end soon if energy cannot be procured. Modern civilization may not last and humans must be serious about changing something.

In Japan we have the term *mottainai*, which represents the attitude of not wasting things and not wasting money. This word summarizes my talk today. So I would like to conclude my talk with this word: *mottainai*.

References

Campbell CJ (2002) Oil Depletion – Updated Through 2001. Available from the following website: http://www.energycrisis.com/campbell/update2002.htm

The Global Commons Institute, GCI (2003) Oil Reserves & Resources, the Depletion Debate. Available from the Contraction & Convergence (C&C) ©®™ website (http://www.gci.org.uk/Rates_Oil_Depletion_&_Climate_Change.html). Recent data is available from the following websites: The Oil Age, Petroleum Analysis Centre (http://theoilage.org/); Jean Laherrère, ASPO France (https://aspofrance.org/tag/jean-laherrere/)

Yoshinori Ishii, President, Mottainai Society (NPO); Professor Emeritus, University of Tokyo; and former professor, Toyama University of International Studies, Doctor of Engineering, Research interests: global environmental science, energy and environmental theories, remote sensing, engineering in exploration geophysics.

Dr. Ishii graduated from the Department of Physics, Faculty of Science, University of Tokyo in 1955. After working for oil companies for 16 years, he first assumed an assistant professorship and then a professorship in 1978 at the Faculty of Engineering, University of Tokyo. He joined the National Institute for Environmental Studies, Japan in 1994 (working as 9th president between 1996 and 1998), and worked as a professor at Toyama University of International Studies between 2000 and 2006. He set up an NPO, Mottainai Society, in 2006 and became its president. He has held many posts, including the representative of the Forum of Strategies for Humanity's Future organized by the Engineering Academy of Japan.

His publications include: *Environmental Science for Citizens* (in Japanese), Aichi Shuppan, 2001; *Prosperous Petroleum Age Ends – Where Will Humanity Go?* (in Japanese), Engineering Academy of Japan's Environmental Forum, ed., Maruzen, 2004; *The Peak of Petroleum Has Come – "Japan's Plan B" to Avoid Collapse* (in Japanese), Nikkan Kogyo Shimbun, 2007; and *Oil Peak Triggers Food Crisis* (in Japanese), Nikkan Kogyo Shimbun, 2009.

Websites

http://www.mottainaisociety.org/ (*Mottainai* Society)

The recent state of the author can be found at:

https://oilpeak.exblog.jp/

Chapter 3
Sustainable Use of Energy

Hans-Peter Dürr

I have already visited Nagoya several times, but I have not previously talked about energy. As a nuclear physicist, I am well aware that today is the 60th anniversary of the atomic bombing of Hiroshima. I think that it was the worst tragedy in the history of mankind. As a member of the human race, I believe that nuclear weapons should never be used. We must find a way to overcome these problems.

The question is whether we can design the future of humanity. I think we need to design the future, paying more attention to the environment. Albert Einstein told us that we would have to learn to think in a completely different way and indeed, this is what we must do. We are at present in a major crisis. We have not only to worry about our environment and how to survive, but also realize that the human species is only one part of a huge organism here on earth. Humans must learn to share their creativity with other species in order to survive, because while nature can survive without humans, we cannot survive without nature as we find it here on earth. With this in mind, my lecture today will discuss the sustainable use of energy.

First of all, I would like to discuss sustainability. What does sustainability mean? It concerns how we should live as a community, but more than that, how we should participate in nature as well. I would then like to talk about the problems of energy. The key to achieving sustainability is energy.

3.1 The Earth, the Sun and the Universe

We do not inhabit the entire earth: only its surface. The zone we inhabit stretches from 10 km up into the atmosphere above, to 1 km beneath the surface. And, there are subsystems (Fig. 3.1); there are materials and energy. Every day the sun provides us with energy. The sun's energy is the sole source of all our energy.

Dr. Hans-Peter Dürr †, Director Emeritus, Max Planck Institute for Physics and Astrophysics, Germany; Professor Emeritus, Ludwig-Maximilians-University, Germany.

© The Author(s), under exclusive licence to Springer Nature Switzerland AG 2020   19
Y. Hayashi et al. (eds.), *Balancing Nature and Civilization—Alternative Sustainability Perspectives from Philosophy to Practice*, SpringerBriefs in Environment, Security, Development and Peace 32, https://doi.org/10.1007/978-3-030-39059-4_3

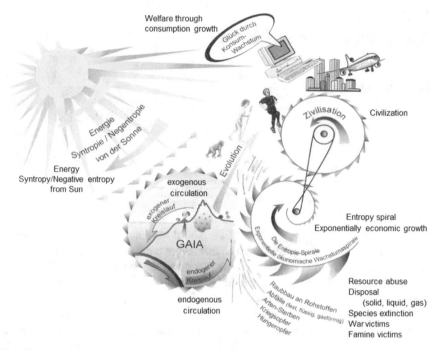

Fig. 3.1 Geo-ecosystem. *Source* The author

Of course we cannot use this energy forever, but we should be able to use solar energy for the next 5–6 billion years until the sun burns out.

The sun is the only source of energy. Other forms of energy are limited, so that we are stealing from the earth. When we survey the earth, we sometimes find new materials, and we are under the illusion that new materials are being born. Looking at the economy, we use materials as well as energy. However, we are just using materials that have been preserved on the earth. For example, when we use copper, it means we are using a limited resource.

All energy comes from the sun. Almost all solar energy is reflected by the earth's surface and emitted back into space (Fig. 3.2). Some energy is conserved by biosystems and plants, but only a little: in the region of 1/10,000th. In other words, we can only make use of energy to that extent.

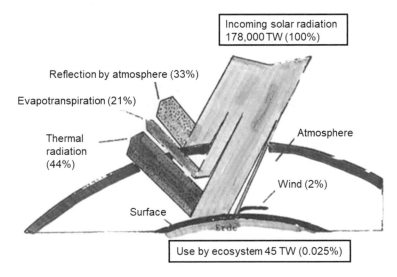

Fig. 3.2 Solar energy. *Source* The author

3.2 Usable Energy: Syntropy

If we are going to use energy, the important point is that we have to store it. Solar energy is a usable energy but we can use only part. The energy that we cannot use is emitted back into space as radiation.

The usable energy is the source not only of energy, but also of syntropy (Fig. 3.3). Syntropy is the opposite of entropy, which is disordered; that is, syntropy is ordered and high-quality energy. The usable energy that we use is syntropy.

3.3 Humans 'Steal' Resources

Humanity uses energy in processes which add value. We have depended upon resources that are underground (Fig. 3.4), resources which are limited. Coal, gas and oil have taken hundreds of centuries to form, but we humans have been using them only in the past one to two centuries. In other words, we are wasting them. They are limited resources, which have stored energy over a long period of time. We use uranium as fuel for nuclear fission. Uranium is also limited. Uranium was produced by exploding supernovae billions of years ago. As for hydrogen, we use the hydrogen that was formed in the Big Bang. It is on the earth, but is also limited.

We live our lives by using limited resources. We can call our society a "Bank robber society" (Fig. 3.5). It means that we do not produce something new, but we open the 'safes' of Nature time after time, taking out and using various resources. Furthermore, we make tools to extract various resources. In this sense, we are like

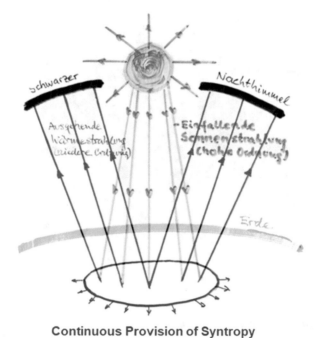

Continuous Provision of Syntropy

**= Primary motor for the evolution of life on Earth and
processes of human value creation**

Fig. 3.3 Sun as source of syntropy. *Source* The author

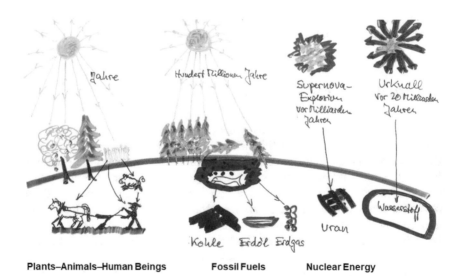

Plants–Animals–Human Beings   Fossil Fuels   Nuclear Energy

Fig. 3.4 Islands of high syntropy on earth. *Source* The author

Bank Robber

Investment
in welding
equipment to
break into the
"safes" of Nature
one after the
other.

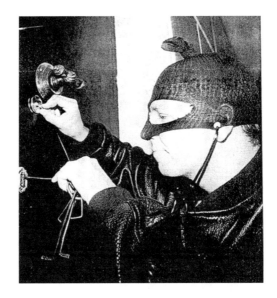

Fig. 3.5 Bank robber. *Source* The author

bank robbers or akin to the Mafia. Since it takes a long time for resources to be produced, it is much easier for us to steal. We have been living by stealing resources from Nature like bank robbers.

3.4 The Unstable Biosystem

In addition to limited resources, there is another difficult problem. It is the biosystem (ecosystem). The biosystem is a system comprising the animals and plants which live on the earth. As we are accustomed to machinery, we tend to interpret the biosystem as a complex machine. In fact, it is not a machine, but an unstable system. This unstable system is kept in balance by homeostasis. Homeostasis requires energy. Energy is required to maintain a good balance. Solar energy is indispensable in maintaining the stability of the biosystem.

This biosystem can be compared to a house of cards (Fig. 3.6). We stand on the house of cards, and think we are on the top of the ecosystem. In fact, we are standing on a fragile house of cards. When the energy of the sun reaches the house, every card tries to keep balance by moving forward and backward, or up and down. The solar energy maintains the stability of the pyramid of life. In order to keep the biosystem stable, energy of 45 TW (45 terawatts; i.e. 45 trillion watt) is required.

Biosystem

Many people think they are the apex of creation and jump around on top of the house of cards. They do not see that cards are collapsing and falling out, and that, hence, their own foundation is seriously endangered.

Photo: Seidel/Weidlich

Fig. 3.6 Biosystem. *Source* Photo editing by Seidel and Weidlich

3.5   Dynamic Sustainability of the Biosystem

If the biosystem were stable, like a mountain, we would be unlikely to fall from the mountain top. However, we are really depending on a house of cards which is very unstable. We have to bear this in mind, and think correctly.

The German word for sustainability, *Nachhaltigkeit,* literally means that nothing will happen if it is conserved (Fig. 3.7). Maintaining a certain state has a static meaning, but this does not well express the sustainability of the biosystem. The

Dynamic Meaning of Sustainability

Ability to sustain in a non-static way

(In German: Nachhaltigkeit connects *nach* ("after") and *halten* ("maintain"))

Guarantee and support for the dynamic process of life

Vitality, productivity, creativity, resilience, robustness

Letting the living be more alive!

"I am life that wills to live
in the midst of life that wills to live."
– Albert Schweitzer

Fig. 3.7 Dynamic meaning of sustainability. *Source* The author

biosystem is dynamic. It has the energy to produce something. It not only maintains a certain state, but has a dynamic character in the sense that it maintains growth and activity in the biosystem.

The Latin root of sustainability means "to make livelier". Observing the Earth's ecosystems, Albert Schweitzer said: „Ich bin Leben, das leben will, inmitten von Leben, das leben will." [I am life that wills to live in the midst of life that wills to live.] This is the philosophy that we need to bear in mind. We are just a small part of the larger ecosystem that surrounds us. This idea is based on the philosophy of what can be called "monsoon Asia". The whole universe is an ecosystem with a very dynamic character. I hope you will understand 'sustainability' as such.

3.6 Creation of Values: Nature Versus Economy

What are the essential issues for sustainability? The issues can be considered on various levels, not only the level of the ecosystem. How can humans co-exist? It is important to see humans as individuals as well as from an economic viewpoint. We live as humanity as a whole. We are not just consumers, but we are creating something. We have to consider three levels: ecological, social, and individual. Humanity is *Homo sapiens,* which originally means wise men, so we have to be wise men in the true sense of the word. We have to be wise not only in terms of consuming, but also in terms of creation. Economy and ecology are not in conflict with each other. Let us consider the creation of economic value, and the creation of natural value. Economics does not create value itself. We are taking the value in nature and using it for our own purposes. We are conducting various economic activities. However, increasing numbers of people do not create, but just cause 'avalanches' in the snow.

Nature has many positive-sum games. These are games in which the total outcome is positive; in other words, one side's gain can also result in the other side's gain, with an overall positive outcome. This is the paradigm of living creatures, and value is created. However, there are also games in which the total outcome is non-positive. In these zero-sum games or negative-sum games we become losers. An increase in entropy means that the future will become chaotic. If we could take the snow from an avalanche back to the top of the mountain, it would bring about creativity, but what we are doing is like causing an avalanche. That is the creation of value in the economy.

Thus the creation of value in the economy is entirely different from the creation of value in nature. We are now destroying value. When we use energy, for example, we use the resource called coal, and convert it to heat energy; and that heat energy becomes wasteful and useless energy. By burning coal, we convert it to heat energy that we cannot use, and emit CO_2. We have to change the way we think about this.

3.7 Limited Resources and Energy

The economic system was once very small compared to the entire earth or the earth's surface (Fig. 3.8). Human activity was just a small part of nature. If we extracted resources from the earth, there were still plenty of resources left. If we produced some waste, we did not have to worry about it. However, what about now? Economies have become very large. We now have to consider where we extract resources from, and think about our future. We need to realize that the

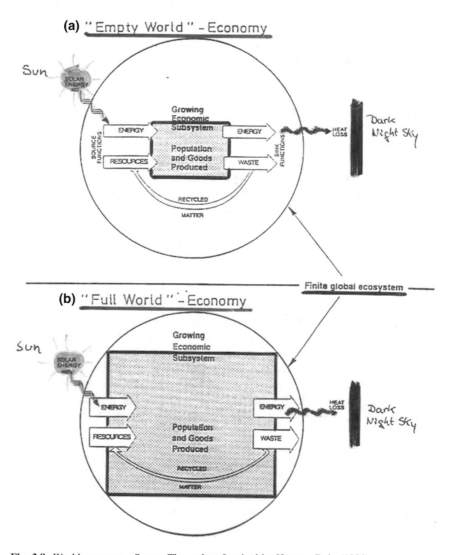

Fig. 3.8 World economy. *Source* The author. Inspired by Herman Daly (1990)

resources we once thought of as limitless are now limited. When we dispose of waste, we are now in the position that we cannot throw it away as garbage.

Our political systems cannot yet deal with these problems. We have parliaments which are elected only by a majority of the people currently living. Laws are made on the basis that what was right in the past is still right. But who is thinking about the future? We need new forms of political assembly that work for our future. We have to consider what will become of our children's and grandchildren's generations.

3.8 Energy Slaves

Next, let us discuss energy (Fig. 3.9). Energy is syntropy, and the source of the evolution of life. The primary energy is solar energy. Secondary energy is the accumulated energy of the past, made from organisms such as animals and plants. What we must realize is that we are now using this limited energy for our own purposes, for mere pleasure. If we continue like this, the house of cards that is the biosystem will collapse at some point.

What keeps the house of cards stable is energy. 13 TW are required, which if we convert to kilowatt hours is 13 billion kWh. In terms of manpower (0.1 kWh/person), this is the equivalent of 130 billion energy slaves. At present, there are 6 billion people on the earth, but 130 billion slaves are working to support us.

In these circumstances, we have to realize that sustainability has already been undermined in many fields. We have consumed huge quantities of limited resources and produced enormous amounts of waste. We do not yet know where to dispose of nuclear waste. Nobody yet knows where to store the CO_2 that is being emitted in great quantities.

Key role of energy

Motor for evolution of life
> Primary source: Solar energy – "Free" syntropy source
> Secondary source: Organic life forming food, plants and animals

Motor for industrial development and revolution
> Primary source: Solar energy in the form of wood, water, wind
> Secondary source: Fossil fuels such as coal, oil, natural gas
> Nuclear sources: Fission (and fusion)

But also

Gross energy throughput measure for stress on biosystem
> Anthropogenic primary energy turnover (1992)
> 13 TW = 13 billion kW per hour = 130 billion energy-slaves
> (Manpower-equivalent: 1 energy-slave (100 W = 0.1 kWh))
>
> (Biosystem dynamically stabilized by about 40–50 TW solar radiation energy)

Fig. 3.9 Key role of energy. *Source* The author

As a result, the ecosystem on the surface of the earth is very unstable. Concrete problems such as the excessive accumulation of CO_2, as well as the resource scarcity that was earlier compared to a bank robber society have made it necessary for us to change our behavior. We now consume fossil fuels with a carbon conversion weight of 6.6 GtC (GtC: Gigatonnes of carbon), and emit CO_2 into the atmosphere. Unless we drastically reduce this amount, a huge quantity of CO_2 will accumulate in the atmosphere and the climate will become highly unstable.

3.9   Unequal Distribution

The consumption of primary energy was 13 TW (130 billion energy slaves) in 1992 (Fig. 3.9). The amount has probably not changed very much since then. I think we can learn a lot from this.

First of all, we need to consider who is exploiting these 130 billion energy slaves. This is because there is a very unequal distribution. A single American currently exploits 110 energy slaves a year, so a family of four exploits 440 energy slaves per year (Fig. 3.10). In other words, 440 slaves are roaming about in a typical family's house. In Europe, 60 energy slaves are exploited per person, while Japan exploits 50 energy slaves per person – about the same level as Switzerland (Fig. 3.11). On the other hand, China exploits 8 energy slaves per person and Bangladesh less than 1. This is a highly unequal distribution of energy usage.

People ask how many humans can fit on this earth before our biosystem collapses? Is 6 billion too many? 20 billion? The answer is that it doesn't matter at all. You have to ask how many additional energy slaves are required. You need "birth control" of energy slaves. We have 130 billion energy slaves. If you get a mid-range car, you may not realize that you have 250 energy slaves underneath your foot when you take a letter to the mailbox.

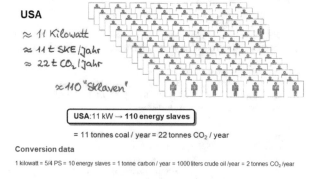

Fig. 3.10 Personal energy consumption (USA). *Source* The author

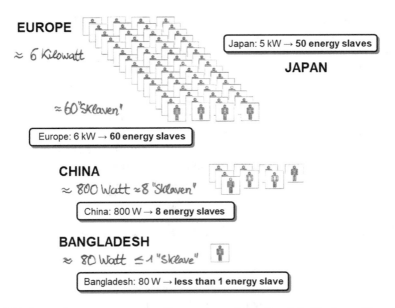

Fig. 3.11 Personal energy consumption (Europe, Japan, China, Bangladesh). *Source* The author

3.10 Permissible Energy Usage

How many energy slaves does the biosystem allow humanity to use? We calculated this and made the list shown in Fig. 3.12.

The required number of energy slaves was calculated for each of the following resource bases: carbon, hydrogen, nuclear energy, biomass and water power. In the case of the carbon-based resources, about 10 energy slaves are produced. We add

Carbon Dioxide production   Global (1990)

		C - portion 61%	6.1 TW	→ 5.7 GtC/year
Fossil fuels	10.0 TW	H² - portion 39%	3.9 TW	→ 0

Each C energy slave produces 5.7 / 61 tonnes C / year = 93 kgC/year = 338 kg CO_2/year

Energy slaves per person

(130m energy slaves per 6m people)

C energy slaves	47 %	10.2	950 kg C/year
H₂ energy slaves	30 %	6.5	
Nuclear energy slaves	5 %	1.0	
Biomass energy slaves	12 %	2.7	
Water energy slaves	6 %	1.3	
Total number of e-slaves/person		21.7	produce 950 kg C/year

Stabilisation of climate requires the maximum load on global average to be less than 250 kgC/year/person or 3 C energy slaves/person.

Fig. 3.12 Carbon dioxide production and energy slaves per person. *Source* The author

up the number for each resource, and the total is obtained. As shown in Fig. 3.12, we now live exploiting about 22 energy slaves per person. On the other hand, we came to the conclusion that no more than 90 billion energy slaves (9 TW) should be exploited in the whole world in order to maintain the current biosystem, so only 15 energy slaves are allowed per person. In other words, energy usage should be less than 1.5 kWh per person to maintain the earth.

Considering the current energy usage of 130 billion energy slaves (118 billion slaves if we consider the unused ones), the usage already exceeds the permissible upper limit of 90 billion. Supposing the energy usage of advanced countries increases every year by 2.5%, and the poor countries maintain current usage, total world usage will reach as many as 300 billion energy slaves, and greatly exceed the upper limit.

3.11 The Road We Must Travel: Spirituality-Based Cooperation

Now, how do we overcome this dilemma? We have taken a lot from nature in order to maintain our current economic activity. This economic activity is a zero- or negative-sum game. In a highly diversified society, do we not have to build a society in which people can cooperate with each other to a great extent? We now need to cooperate by abandoning various forms of competition in the world. The Latin root of 'cooperate' means to work together. Nowadays the word often seems to be used with a different meaning. I think we should go back to the original Latin meaning and build a society on a higher level in which we cooperate by working together.

Of course, I may be talking about an ideal case, but I would like you to at least consider my opinion as one point of view. Physics, which is my specialty, has progressed to a remarkable extent. In present-day physics matter is not regarded simply as matter. Matter does not exist only as matter, but is also connected with various other forms of matter. We are now at a time when we should review how humans have evolved. I would like to suggest that matter and life – as well as spirit – are not independent of each other, but must be dealt with holistically. I believe we should understand that we are born to cooperate with each other. It may not be possible to argue that the human race was, from its earliest days, cooperative, and avoided strife. However, viewing our species in the way I have indicated shows that we need to create a society in which we can cooperate with each other. We need to live with a high degree of spirituality.

3.12 A Satisfying Life

The limitation of resources is not the main problem because we cannot continue with this waste of energy. We will have to operate in different ways. One is efficiency. For example, in Japan, you have 50 energy slaves, but half of these energy slaves are due to inefficiency, and can be sent away. Then there is still the question of how to get down from 25 energy slaves to 15 energy slaves. This is a problem that has to be managed, but it is not completely impossible. These 15 energy slaves that I'm talking about represent the lifestyle of the Swiss in 1969. This is what you have to do.

The second way is sufficiency. We have to ask the question: how much is enough? If you say you can never have enough, you are heading for a dead end. But don't be afraid: this life can be a very happy life and that is what we have to deal with. We have to look at lifestyles which are within these limits of 15 energy slaves. We can become very creative. You can try it out in this way: we have made a sheet which we send to people to calculate how many energy slaves they are really employing.

I can ask all of you. You here in Japan have, on average, 50 energy slaves and you may find it interesting and surprising what those energy slaves are doing. They are wasting most of their time. If each person realizes how they are using more than 15 energy slaves, they will find a way to adopt a lifestyle which fits into that. I will tell you, there are many things which you can do to survive that are really fun, and that is the direction in which we have to go.

Reference

Daly HE (1990) Towards Some Operational Principles of Sustainable Development, *Ecological Economics* 2:1–6. See also Daly HE (1996) *Beyond Growth: The Economics of Sustainable Development*. Beacon Press, Boston, p 46

Hans-Peter Dürr The late *Hans-Peter Dürr,* Director Emeritus, Max Planck Institute for Physics; and Professor Emeritus, Ludwig-Maximilians-University Munich, Doctor of Physics, Research interests: nuclear physics.

Born in Stuttgart, Germany in 1929, the late Professor Dürr graduated from the University of Stuttgart in 1953, and was awarded a doctoral degree from the University of California, Berkeley in 1956. He worked at Max Planck Institute for Astrophysics, and, as a co-researcher of Werner Karl Heisenberg, contributed to creating the theory of quantum mechanics and a unified field theory. In 1969, he was appointed to a professorship at Ludwig-Maximilians-University Munich. He started to work also at Werner Heisenberg Institute in 1971 (later Max Planck Institute for Physics). He was awarded various highly-regarded prizes including the Right Livelihood Award in 1987. He became a full member of the Club of Rome in 1991, and theorized nuclear threats and the sustainability of Earth Gaia holistically to bring them to people's attention. He edited the Potsdam Manifesto 2005 (50 years after the Russell-Einstein Manifesto 1955). He died in 2014. His publications include: (co-editor) *Werner Heisenberg. Collected Works,* 9 volumes, Piper and Springer, 1985–1993; and (co-editor and co-author) *What Is Life?* World Scientific Publishing, 2002.

Chapter 4
Sustainability from the Perspective of Environmental Archaeology

Yoshinori Yasuda

4.1 Pollen Analysis

I have been studying fossilized pollen. Pollen is so tiny that it cannot be observed with the naked eye, but it has a chemically strong membrane. When pollen accumulates on the bed of a lake or wetland, it will survive there for tens of thousands of years without decaying. My research concerns extracting fossilized pollen from earth to reconstruct a past environment, as well as understanding how forests or the climate changed. Today I would like to discuss how sustainability and the results of pollen analysis are related.

You may have heard of Easter Island, and Fig. 4.1 shows fossilized pollen from palm trees which was found on the island. Hardly any trees can be found on the mountain on Easter Island today, but the discovery of palm tree pollen reveals that the island once had a large palm tree forest.

4.2 What We Eat Changes Our Fate

As my first topic, I would like to discuss what humans eat. Nowadays it is often felt that many children easily lose self-control, and this has brought the effect of diet on children's emotions to people's attention. It should also be noted that there is a close connection between human diet and the effect on Nature. We eat to live. Because humans must live, they need to eat. We should realize that eating has a big impact on Nature (Fig. 4.2).

On the other hand, in the western part of the Eurasian Continent around the Mediterranean Sea and western Asia, there is not enough rainfall. There is half as

Yoshinori Yasuda, Director, Research Center for Pan-Pacific Civilizations, Ritsumeikan University, Japan; Professor Emeritus, International Research Center for Japanese Studies, Japan.

© The Author(s), under exclusive licence to Springer Nature Switzerland AG 2020
Y. Hayashi et al. (eds.), *Balancing Nature and Civilization—Alternative Sustainability Perspectives from Philosophy to Practice*, SpringerBriefs in Environment, Security, Development and Peace 32, https://doi.org/10.1007/978-3-030-39059-4_4

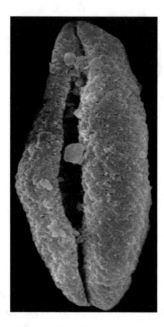

Fig. 4.1 Fossil palm pollen from Easter Island. *Source* The author

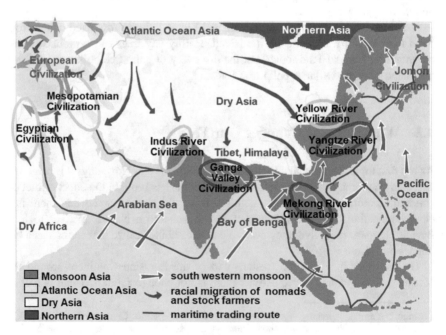

Fig. 4.2 Climate division of Eurasia. *Source* Yasuda (2002: 51). Reprinted/adapted with permission from Chuokoron-Shinsha, Inc. © 2002 Chuokoron-Shinsha, Inc./Yoshinori Yasuda

much as in Japan – no more than 500 mm a year. Moreover, as it rains mainly in winter in these areas, people there grow winter crops such as wheat and barley to make bread and obtain protein from the meat of sheep or goats. They also make butter and cheese from the milk of sheep or goats. This lifestyle started about 15,000 years ago.

When the global climate shifted from the ice age to an interglacial period, people in the west and east of the Eurasian Continent chose two different forms of diet. One lifestyle, which is ours in Japan, is based on a diet of rice, fish and miso soup, whereas the other is a diet of bread, milk and meat. These lifestyles have continued for 10,000 years, and how has the environment of the globe changed over the period?

4.3 Wheat and Livestock Farmers Devastated Forests

What did the people with a diet of bread, milk and meat do over the past 10,000 years? It was in Mesopotamia that this dietary pattern was established. The landscapes that comprised Mesopotamia are today without forests. Figure 4.3 shows the results of an analysis of pollens from the Ghab Valley, Syria. Wheat farming started there 12,000 years ago, and surprisingly it led to serious damage to forests 10,000 years ago. By about 5,000 years ago, almost all of the forests in the surrounding areas had disappeared. This is truly surprising.

Why did the people living in this region have to destroy the forest? The cause of the destruction was, ultimately, protein, which is essential for human life. The people kept sheep and goats for their milk, to make butter and cheese, and for meat to eat. This dietary pattern resulted in the destruction of the forest. The people who lived here did not hate the forest; in fact, they worshiped forest gods. However, sheep and goats eat grass and young buds from morning to night. Even if they are destroyed by humans, forests will grow again naturally – so long as they are left alone. However sheep and goats consume the buds essential to the regrowth of forests. For this reason, as the number of sheep or goats increases, the forests there will be completely destroyed. Figure 4.4 shows the current landscape of the Ghab Valley, from which pollen was collected for analysis. The valley once had a rich forest of deciduous oaks, which were all destroyed, and today there is neither soil in the surface layer nor water.

After the fall of Mesopotamia, civilizations emerged around the Mediterranean, and especially Greece. Greece is now well known for its bare mountains. One might think forests had never grown in Greece, but an analysis of pollen reveals that the land was once abundant with forests. When the Greek civilization flourished, Attis and other gods of vegetation were worshiped, and the country was covered with a rich forest of deciduous oak trees, as shown in Fig. 4.5, which is the result of analysis of pollen from around the former site of Lake Copais in Boeotia. The forest has been completely destroyed.

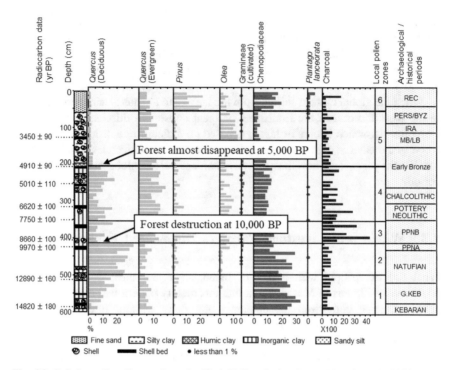

Fig. 4.3 Relative pollen diagram from the Ghab Valley, Syria. *Source* Yasuda et al. (2000: 131). Reprinted/adapted with permission from Elsevier Science Ltd. and INQUA. © 2000 Elsevier Science Ltd. and INQUA

Fig. 4.4 Bald mountain around Ghab Valley. Forests and water completely disappeared from the West Asian Fertile Crescent. *Source* The author

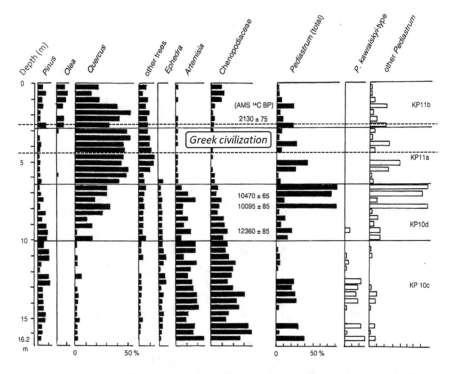

Fig. 4.5 Selected pollen diagram on the top 16.2 m part of a 120 m core from Lake Copais, Southeast Greece. *Source* Okuda et al. (1997: 109). © The Author(s)

In addition, the destruction of the forest caused the sea to be less fertile; nutrients stopped flowing from the forest into the sea. Figure 4.6 shows a landscape from a Greek island. No seaweed can be found on beautiful Aegean beaches; in its place the bodies of Westerners bask like sea-lions. The Aegean has become so infertile that you will never be stung by jellyfish if you swim in it. It is now clear how much nutrient the vanished forest once delivered. For example, the nutrient iron was of considerable importance: it flowed from the forest and nourished plankton, which were in turn fed on by fish. In a passage from Homer's epic, people on an island discuss the most delicious food. I supposed that the people of the island would answer that fish was the most delicious, but instead they asserted that it was lamb. This suggests that ancient Greeks prized lamb above other foods. Even island-dwellers preferred lamb.

I argue that the destruction of the island's forest was a result of the mainland mindset of its inhabitants. Since people had traveled an expansive mainland area, grazing their livestock, they considered resources limitless. Japanese people, on the other hand, living on smaller islands with limited land area would be more naturally aware of the likelihood that grazing livestock might eradicate forests. Greek civilization is considered the root of European, and thus modern, civilization. That civilization was established by people who grazed their livestock on islands.

Fig. 4.6 Greek islands and the blue Aegean Sea. Destruction of forest stopped flow of nutriments so sea became barren. *Source* The author

Wheat and livestock farmers who keep sheep and goats cope with a river basin by establishing pastures and fields on a plateau. They grow wheat there and they put livestock out to pasture on the mountain. What happens next is that sheep and goats keep eating grass until the forests are destroyed. As a method of land usage it might strike us as comparable to using a well as a toilet. How could they have chosen such an absurd way of using land? I am speaking frankly because there are very few people from Europe in the audience today; otherwise, I would talk about this in a milder manner. What happens when livestock grazes near water sources? Every day, they eat and excrete, which contaminates the groundwater.

Since the water in the surface layer is contaminated in almost all parts of Europe, neither the groundwater nor surface water in many parts of Europe is drinkable. On the other hand, the surface water in Japan is still drinkable today because livestock is not put out to pasture. In addition, forests in Europe have been destroyed and nutrients no longer flow from forests to the sea. As a result, the number of fish decreased and the Mediterranean civilization came to an end.

Where did the center of civilization go next? It went beyond the Alps to mainland Europe in pursuit of forests, and by as early as the 17th century, 90% of forests in Britain, 70% in Germany and 90% in Switzerland, for example, had been destroyed. As a result, the forests are gone, as Professor Ishii and Professor Dürr have told us. There were no longer energy sources. This led to the use of fossil fuel, and to the Industrial Revolution.

In this context, modern civilization emerged from the civilization based on wheat farming and livestock farming. What sort of civilization is this? It exploits Nature one-sidedly. Once a place was exploited, it became a desert and civilization moved to another place; this is how it was. So when forests in Europe were destroyed on a massive scale, the people there faced a serious challenge.

Professor Hiroyuki Kitagawa of Nagoya University, my co-researcher, researches climate change. By analyzing the rings of the long-lived Japanese cedar trees on Yakushima (Fig. 4.7) he has found that the climate in the Little Ice Age deteriorated in the Edo period (17th to mid-19th century). In the Little Ice Age, many people in Europe facing challenging circumstances moved to America. What did the migrants do there? The left part of Fig. 4.8 shows the distribution of forests in the United States in 1620, when there was no significant Anglo-Saxon population there. After migration to America, and the introduction of domestic sheep and goats, 80% of the forests in the United States were destroyed in only 300 years (Fig. 4.8, right). Not only the forests, but the water cycle of the country was also destroyed, with its attendant massive impact. In the agriculture of the contemporary United States, maize farming on the Great Plains, for example, does not depend on surface water, but entirely on groundwater from the Ogallala Aquifer. This will raise the same problem as petroleum – using underground resources. What if the resources are used up? That is exactly what we have been debating: global environmental issues.

A similar problem has occurred in Asia too. In the area of northern China along the course of the Yellow River, sheep and goats are tended. They are kept by Han Chinese who live on a diet of bread, milk, butter, cheese and meat. This area, the Huangtu Plateau, was once covered with massive forests, which survived until the Qing Dynasty, but have since turned into deserts.

4.4 Rice Farmers and Fishermen Protect Forests and the Water Cycle

A diet of rice and fish has resulted in the protection of the ecosystem of forests and the water cycle. We were not confident about our diet of rice and fish, and that is why we started having bread for breakfast after World War II; eating bread felt stylish. Of course, it is good to eat bread, but we should eat rice too.

Our recent research has revealed that rice cultivation started as early as 14,000 years ago. The Yangtze civilization in China, which was supported by rice

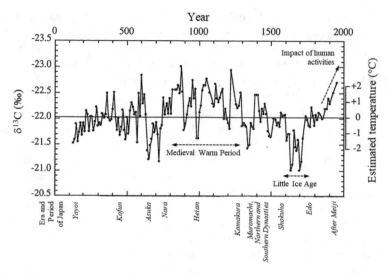

Fig. 4.7 Climate change restored from the δ^{13}C record of a giant cedar tree from Yakushima Island, southern Japan. *Source* Kitagawa (1995: 50). Reprinted/adapted with permission from Asakura Publishing Co., Ltd. © 1995

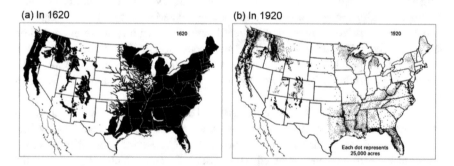

Fig. 4.8 Distribution of forest in USA. **a** In 1620, **b** In 1920. *Source* Goudie (1993: 45). Reprinted with permission from Blackwell Publishers. © Goudie 1981, 1986, 1990, 1993. See also Yasuda (2002: 7)

farming, was in existence as early as 6,000 years ago. Figure 4.9 shows the landscape of the Yuchanyan cave site in Henan Province, China. Rice cultivation started in this area 14,000 years ago. What became of this rice farming area, which lies to the south of the basin along the lower course of the Chang Jiang (Yangtze) River? Forests and the water cycle are well preserved in this area. The area's landscape is really beautiful. A variety of organisms, such as water beetles, water striders, aquatic weed and fish, live in paddy fields; biodiversity is preserved there. A recent study has also revealed that paddy fields purify groundwater.

Fig. 4.9 Yuchanyan in Hunan Province, China. Rice cultivation began around 14,000 years BP.
Source The author

Terraced paddy fields in Fig. 4.10 illustrate the most important point. For those who keep sheep and goats, such a steep slope can only be used as pasture. What happens then? The sheep and goats eat up the grass, and the soil of the resulting grassless land is eroded, becoming barren.

What did the rice farmers and fishermen do with land that would become barren if left unmaintained? They parceled it out among family members: this field is my grandfather's, this is my father's. The slope is so steep that machines could not be used. So rice farmers and fishermen made the effort to cultivate the steep slope, putting their energies to work in making barren earth fertile. As Professor Dürr has pointed out, a very important issue in the 21st century is finding ways to save energy. Rice farmers and fishermen have implemented an energy-saving practice: they put their energy into the land to make it fertile. This is their wisdom. On the other hand, wheat and livestock farmers take a different approach: they graze livestock on the land, which turns it into a wasteland. Rice-farmers and fishermen put their energy into the land and derive happiness from making the land fertile. This is very important in the 21st century. It is not new machinery or energy that is used to make wasteland fertile. Reviving Nature using the strength of one's own limbs does not use up energy sources. Perhaps such work may mean people have to eat a little more, but it makes hardly any difference in the amount of energy consumed. This is how rice farmers and fishermen go about using land.

Fig. 4.10 Terrace paddy field in Guizhou Province, China. Rice farmers devote their energy to creating fertile land. *Source* The author

4.5 Mountain-Worshiping Mindset

Mountains are not toilets. They are holy places. They should have forests. They should have *kami*. They inspire awe in us. Most of you here see mountains and feel something divine, but people from Europe regard mountains as no more than places for playing sport. Mountains are challenges to overcome, things to surmount. However, we Japanese all look at Mount Fuji and find it awe-inspiring. We consider it divine because this mountain's water source is the fountain of life for rice farmers and fishermen, which is what we are. No one should set up toilets or graze livestock there. Water from the mountain irrigates rice fields, and the water from the rice fields carries nutrients to the sea. The nutrient nourishes fish, which feed us. Water evaporates from the sea and falls on the mountain as rain. As long as this water cycle continues, the basic level of our lifestyle is guaranteed. If petroleum or chemical fertilizers become unavailable, fallen leaves and branches can be used as fertilizer. Undergrowth was once used as a fertilizer. As long as the water cycle is well maintained and continues, we can live forever on Earth.

4.6 Nurturing Altruism and Benevolence

It is also important to share this beautiful water with everyone. When growing rice in irrigated fields, you bring water to your paddies. Using up all the water brought to your paddies will cause problems for other people. Therefore, after you have brought water to your fields, you have to clean the water and return it for other people to use. In a society where rice is farmed in irrigated fields, people need to consider the interests of others. This is different from a society based on wheat and livestock farming. In this type of society, the farmer claims the water on his farmland as his own because water falls from the sky as rain. For this reason, in a society based on wheat and livestock farming, a farmer will try to expand his farm. As his flock of sheep increases in number, the farmer gets richer. He tries to invade other groups' territory to expand his own, and keep more sheep. This is how to increase wealth. On the other hand, a society based on rice farming in irrigated fields cannot function like this because it operates within the restriction of water. Now, in the 21st century, the behavior of all humans on Earth is restricted by water. The time is coming when the fate not only of Japan and its rice farmers, but also of wheat and livestock farmers depends on water.

The future of rice farmers and fishermen is not promising, either. For example, at Owase Bay in Mie Prefecture in central Japan, the underwater 'forest' on the bay's seabed has already gone. It is in the places where a variety of seaweeds, such as kelps and sargassum, grow, that fish lay their eggs. Such underwater seaweed forests are rapidly disappearing. One of the reasons for this phenomenon – underwater deforestation – is understood to be global warming, but this has not yet been fully established. Our attention has been directed only to deforestation on land. While we are concerned about the depletion of tropical forests on land, underwater forests are facing a critical situation out of sight. As the places where they lay their eggs disappear, fish will decrease in number. And of course we depend on fish as an important source of protein. The construction of the Three Gorges Dam (on the Yangtze River, China) prevents nutrients from flowing from the Yangtze River to the East China Sea. Some estimates say that catches in the East China Sea will decrease to a fifth or less their current levels in five years. This means that we cannot be sure of our survival in the 21st century even if we eat fish. Aquaculture – how to rear fish – will also be an important key to successful survival.

4.7 Connecting the Chains of Life

We live on this small Planet Earth. Japanese astronaut Soichi Noguchi is flying in the space shuttle now. Some may say the space shuttle is testament to great technological advancements, but in reality, we have to live on Earth – a tiny island in the vast universe; we have no choice but to live on this planet. Only a few people could survive on spaceships, and even in a hundred years that would still be a very limited number. The vast majority of people will have to live on this planet, which is a teeny tiny island of life in the incalculable immensity of the universe. It should be noted too that we can use only 2.5% of all water on Earth. If that is the case, how should we live? The most important thing is the mentality of islanders, which we have developed in our island country. 'Mentality' may sound negative, so I have decided to call it "island wisdom" in English. When I was asked what to call the mentality of the people on the European Continent in English, I decided to call it "continental mentality" to make it sound a little inferior. Island wisdom – it sounds nice. An island mentality has been considered worthless, but once we regard Earth as an island, the ways of thinking developed on islands will play a very important role in creating sustainable civilized society. However, when we compare the mentalities developed on islands on the Pacific Rim, such as Japan, New Zealand, Tahiti and Easter Island, not all of them have developed such positive islander wisdom.

4.8   Forest- and Self-destructive Moai Civilization

Let's discuss Easter Island, for example. There are many huge *moai* on Easter Island (Fig. 4.11). The island's civilization, which produced *moai* as tall as 20 meters, destroyed its forests and finally itself. Why didn't the people there try to preserve the limited resources available from the island? It must have been evident that, if the forests were gone, soil erosion would follow and the amount of available food would gradually diminish. So why did they fail to prevent their civilization from collapsing into chaos? The *moai* offer an explanation. The *moai* built in earlier times are small, but as time passed toward the end of the civilization, bigger and bigger *moai* were built. Resources ran out and food became unavailable. Why didn't the people devote their energy to increasing the amount of food and planting trees, instead of building *moai?*

I have thought about the reasons and come to a conclusion: all of the *moai* are male. Male society is dominated by competition and conflict, which cannot prevent societal destruction in a civilization from spiraling out of control. Civilization will fail in such male-dominated societies. Easter Island turned into a wilderness after the destruction of the forests there. The people could have prevented Easter Island from deteriorating before the island reached its present state, but they kept producing giant *moai*. I believe that the failure of the civilization here is largely attributable to the male-dominated tribal society.

On the other hand, Tahiti, which is also in the southern Pacific, is still covered with verdant forest. What is the stature which stands there (Fig. 4.12)? It is an Earth goddess, a stature of a female. This suggests that a society in which women can play active roles can conserve Nature. Japan used to be such a society.

Before dealing with Japan in more detail I would like to discuss New Zealand. The left side of Fig. 4.13 shows the distribution of forests in the 7th century. In the 7th century, 90% of New Zealand was covered with forests, not because no one lived there, but because the Maori who lived there protected the forests. However, after 1880, European immigrants brought sheep and goats to the country, and 40% of the forests were destroyed in as little as 20 years, between 1880 and 1900. Europeans call it "great work." This is still continuing today (Fig. 4.13, right). As long as contemporary civilization continues to destroy forests in a similar manner, we will end up crashing the Earth.

Fig. 4.11 Moai statue on Easter Island (El Gigante). *Source* The author

Fig. 4.12 Mother goddess in Tahiti. *Source* The author

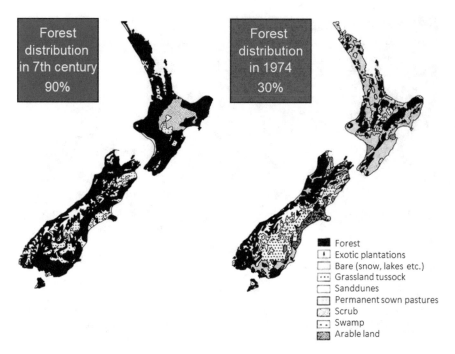

Fig. 4.13 The forests of New Zealand were destroyed after the arrival of Europeans. Left: Forest distribution in the 7th century AD. Right: Forest distribution in 1974. *Source* Goudie (1993: 33). Reprinted/adapted with permission from Blackwell Publishers. © Goudie 1981, 1986, 1990, 1993

What should we do to change it? Bahn/Flenley (1992: 215) described a model showing how the forests decreased on Easter Island. The forests on the island decreased sharply as the population grew. In about the 9th century, the limit of wealth was exceeded. The population continued to grow and immediately after it reached 10,000, the civilization collapsed and the population was reduced by more than half. This is the Easter Island model. If we compare Earth to an island, like Easter Island, in the universe, we can predict that a potential 21st century would follow this model.

Figure 4.14 is a model of the relationship between resources and population, prepared by Donella H. Meadows and other researchers from the Club of Rome, to which Professor Dürr belongs. These researchers estimate that the world population will reach about 10 billion around 2050 and start to decrease slowly to about 8 billion after that. If we apply the model of Easter Island, the year 2020 will be the point where the resources and population will intersect – the peak of wealth. It was the Club of Rome that estimated in 1972 that wealth would reach its peak in 2020, and the club's estimate has proved accurate. Even after wealth reaches its peak, the population will continue growing and reach 9–10 billion in 2050. Contemporary civilization will collapse soon after that. While Dr. Meadows drew a mildly declining population curve, I predict that the population could sharply decrease to a

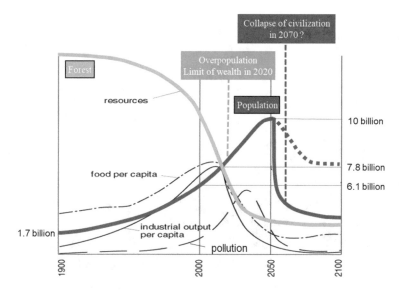

Fig. 4.14 Collapse of present civilization. *Source* Meadows et al. (1972: 124) CC BY-NC; Adapted and rewritten by Yasuda (2005: 89). Reprinted/readapted with permission from Wedge Inc. © Yasuda, Kobayama, Matsui

half or less. A new prediction published in 2004 by Dr. Meadows and other researchers showed that a catastrophe with a drastic reduction in population would occur in about 2070 (Meadows et al., 2004: 14).

4.9 Japan's Model of Rice Farmers and Fishermen

How should we overcome the crisis in the 21st century? Our beautiful Japanese archipelago can be a model solution. Japan has forests, plains and sea, and its nature and environment are well maintained by the forests and water cycle. No other country in the whole world is as beautiful as Japan. Figure 4.15 is a familiar scene: Futamigaura in Ise City. The sun rises between Meotoiwa, or the Married Couple Rocks. We have worshiped both the Sun and the Sea, but which is the main deity? It is Amaterasu – the Goddess of the Sun. We have worshiped a female deity.

If you visit Omiwa Shrine in Nara Prefecture and look for treasure there, you will find nothing. When we open a miniature shine in its prayer hall, we find just a thin wooden talisman. What on earth do we worship? It is the forest and the mountain behind this shrine's prayer hall that we worship. Why do we worship the forest and the mountain? It is because we pray that all the creatures in the forest, all forms of life, and all living things will live forever, and the wonder of life will last too. No treasure is housed in the prayer hall. Mount Miwa rises behind the shrine, and we pray for the mountain. There are trees and woods on the mountain in which

Fig. 4.15 Futamigaura in Ise City, Mie, Japan. The sun rises between the Married Couple Rocks. The larger rock has a small *torii* at its peak. *Source* The author

animals and insects live. Shintoism prays that the order of this world, in which the creatures live, will last and that the cycle of life will continue forever.

4.10 Forms of Love to Save Earth and Humanity

Finally, what sort of solution might save Earth? I wrote that forms of love could be a solution. Professor Yoshitsugu Hayashi, the chairperson of this symposium, phoned me to protest that it could be unsuitable to discuss forms of love at a high-level institution like Nagoya University. So I deleted four or five lines from the handout for the symposium. However, at yesterday's party I told Professor Hayashi about the deletion and he gave me his permission, so I am now able to offer my view to you.

In the 21st century, forms of love play a key role in saving Earth and humanity. The forms of love in Confucianism cover filial piety to parents, as well as love of children, parents and friends, but no more. That is the scope of love in Confucianism. Christianity extended the scope of love to all people, and suggested a wonderful love for all mankind. However, our love – Japanese people's love – is different. We love not only mankind, but also Nature. Japanese people's love, or the love of "monsoon Asians", is broader in its range. Christian love is directed only towards humanity. In the world of Christianity, Nature exists to serve humans. Our love is different. Our love, in the world of polytheism, of Shintoism, of Buddhism, is directed towards not only humans, but also all life on this tiny island in the universe. I call this love *jihi* [benevolence]. Therefore, the upcoming 2005 World Exposition has a nickname in Japanese: *ai/chikyuhaku* ("Love the Earth Expo"). I would like to ask you to consider why it has not been nicknamed simply "Earth Expo" but "Love the Earth Expo." It is because love is going to save Earth and the future of humanity.

Dr. Dennis Meadows, who prepared the Club of Rome's model, made a similar remark. Dr. Meadows also said that humans could be saved ultimately by benevolence, and that nothing but love could do it. I had never met Dr. Meadows until June 2005. At that time, I was invited to the Max Planck Institute, to which Professor Dürr belonged, to attend a symposium to celebrate the 100th anniversary of Einstein's achievements. I wanted to take a subway in Germany, and it cost 2 euros. However, I only had 20-euro bills with me, and was not able to change them. While I was at a loss on the platform, an old white-haired gentleman walked up to me and asked: "Is there anything I can do to help?" I replied: "I want to break these bills, but I can't." "How much do you need?" "2 euros." The gentleman kindly gave me 2 euros. I was impressed by how kind he was, and went to the symposium the next day to find that the gentleman was Dr. Meadows. He argues that it is benevolence that can save humanity, and he put benevolence into practice himself. What is important about this story is that having heard it you should take action yourselves. I will put benevolence into practice too. Now that you are familiar with this anecdote, it is up to you to take action. I hope that you will take the first steps for the preservation of Earth and humanity.

References

Bahn P, Flenley J (1992) *Easter Island, Earth Island.* Thames and Hudson, London

Goudie A (1993) *The human impact on natural environment, 4th edn.* Blackwell, Oxford

Kitagawa H (1995) Yakusugi ni kizamareta rekishijidai no kikohendo (in Japanese) [Climatic change recorded in yakusugi cedar trees]. In Yoshino M, Yasuda Y (eds) *Koza: Bunmei to Kankyo, Vol. 6 Rekishi to kiko* [Study series: Civilizations and climate, vol. 6, History and climate]. Asakura Publishing, Tokyo, pp. 47–55

Meadows DH, Meadows DL, Randers J, Behrens III WW (1972) *The Limits to Growth.* Universe Books, New York

Meadows D, Randers J, Meadows D (2004) *Limits to growth, the 30-year update.* Chelsea Green Publishing, Vermont

Okuda M, Yasuda Y, Setoguchi T (1997) An attempt to subdivide fossil pediastrum from Lake Kopais, Southeast Greece. *Jpn. P. Palynol* 43:107–112

Yasuda Y, Kitagawa H, Nakagawa T (2000) The earliest record of major anthropogenic deforestation in Ghab Valley, Northwest Syria. *Quaternary International* 73/74:127–136

Yasuda Y (2002) *Nihon yo, mori no kankyokokka tare* (in Japanese) [Japan, be a country for environment and forests]. Chuko, Tokyo

Yasuda Y (ed) (2005) *Kyodai saigai no jidai wo ikinuku* (in Japanese) [To survive mega disasters]. Wedge, Tokyo

Yoshinori Yasuda Director, Research Center for Pan-Pacific Civilizations, Ritsumeikan University; Professor Emeritus, International Research Center for Japanese Studies, Doctor of Science, Research interest: Environmental archaeology, geology.

Professor Yasuda was born in Mie Prefecture, Japan, in 1946, and completed a master's degree at Graduate School of Science, Tohoku University in 1972. He taught at the School of Integrated Arts and Sciences, Hiroshima University, and became a professor at the International Research Center for Japanese Studies in 1994. He also worked as a visiting professor at Humboldt University of Berlin in 1996 and as a professor at the Graduate School of Science, Kyoto University in 1997. He is a pioneer in Japan in the new field of environmental archeology, and was awarded Chunichi Bunka-sho prize (by Japanese newspaper company Chunichi Shimbun) in 1996 and the Medal of Honour with purple ribbon (by Japanese government) in 2007.

His recent publications include: *Civilizations' perspectives on environmental history* (in Japanese), Chuokoron Shinsha, 2004; *Introduction to Environmental Archeology – 20,000 years of Japan's Natural and Environmental History* (in Japanese), Yosensha Publishing, 2007; *Civilization of Rice-farmers and Fishermen – From Yangtze Civilization to Yayoi Culture* (in Japanese), Yuzankaku, 2009; and *Mountains Are Battling against Market Fundamentalism* (in Japanese), Toyo Keizai, 2009.

Chapter 5
Re-evaluating the Traditional Japanese Perspective on Nature and Ethics

Minoru Kawada

What does the Japanese traditional perspective on nature and ethics mean in terms of environmental conservation and humans coexisting with nature? This is my theme today. I will base my discussion of these points on arguments developed by Kunio Yanagita (Fig. 5.1), one of the leading intellects of modern Japan. Many of you already know of Yanagita, who is well-known as a folklorist and for his achievements in revealing the reality of the everyday culture of ordinary Japanese, both historical and contemporary. He was born in 1875 and died in 1962.

Fig. 5.1 Kunio Yanagita.
Source Kawada (1997)

Minoru Kawada, Professor Emeritus, Graduate School of Environmental Studies, Nagoya University, Japan.

© The Author(s), under exclusive licence to Springer Nature Switzerland AG 2020   51
Y. Hayashi et al. (eds.), *Balancing Nature and Civilization—Alternative Sustainability Perspectives from Philosophy to Practice*, SpringerBriefs in Environment, Security, Development and Peace 32, https://doi.org/10.1007/978-3-030-39059-4_5

5.1   Introduction: The Japanese Sphere of Life

I will start my report by picking up the threads of Professor Yasuda's contribution. In Japan, quite extensive woodlands still remained. To be precise, forest comprises 68% of Japan's land area, the second highest in the world. As you know, Japan is a highly industrialized and densely populated country. How can such a high proportion of forest preserved in a country which is so industrialized and heavily populated?

It is said that Japan has the highest level of primatology in the world. This is related to maintaining the high proportion of forested land. I heard that when a unique piece of research about monkeys in Japan was published by a research group with core members from the Japan Monkey Center at Inuyama and the Primate Research Institute, Kyoto University, it had a big impact in the world. This is because it was the first research on monkeys to introduce an individual identification system. Before then, the method of monkey research was only to observe the activities of monkey troupes. However, the Japanese primatology research group not only observed troupes, but identified each monkey in a troupe and its activities. Other research groups in advanced countries had been unable to identify individuals in a troupe. The Japanese researchers, however, were able to identify individual monkeys and observe not only troupes, but how each monkey behaved. Thanks to this research, Japanese primatology shot to the top level in the world at a stroke.

The question is, why were Japanese primatologists able to identify individuals? In Japan, the human habitat in a broad sense, and the area of wild monkey troupes' behavior interact. However, this is not the case in any of the advanced nations in which monkey research has been making progress: it is only in Japan. In a sense, therefore, in Japan we are in contact with wild animals in our everyday lives. This enables Japanese researchers to identify individual monkeys. That is the basis for such special powers of observation. Why, then, is there this overlap between the habitats of wild animals and humans in Japan?

These are very interesting facts to consider regarding environmental issues. With these points in my mind, I will consider environmental problems from an aspect of the Japanese perspective on nature and ethics, based on Kunio Yanagita's arguments.

There is a view that the issue of environmental conservation is a problem of policy rather than of people's perspective on nature and ethics. However, even if the policy problem comes first, it requires the support of ordinary people to determine and implement policy. Thus, the perspectives on nature and ethics held by ordinary people are highly significant.

5.2 Perspective on Nature and Ethics for Protecting the Environment

The following three points are often raised in connection with environmental conservation, perspective on nature and ethics to support coexistence between people and nature. Firstly, it is not only human beings whose existence has value; nature too has inherent value. Nature does not just exist for human beings. This came up in Prof. Yasuda's presentation, too. Secondly, the current generation must not limit future generations' chances of enjoying fulfilling lives. This means those of us who are alive at present should not limit future generations through our daily lifestyle and other needs. This is the issue of intergenerational ethics. Please remember this issue, as it is my key point. Thirdly, the natural ecosystem is finite and we should consciously prioritize its cycle of regrowth over other considerations. How have the Japanese traditionally thought about these points?

I mentioned in the introduction as aspects of Japan which should be appreciated, the proportion of forested land, and primatology. Some people may feel this is rather a one-sided view. As you may know, from the late 1960s to the early 1970s, Japan saw significant degradation of the environment and serious pollution: Minamata disease, Itai-Itai disease, air pollution in Kawasaki and Yokkaichi, and so on. I believe some of you may think such negative aspects are missing from this discussion. I will talk about that later in connection with my third point.

5.3 Traditional Intergenerational Ethics

First, I will review the first and the second points, starting with the second, the issue of intergenerational ethics. According to Yanagita, Japanese tradition is distinguished from other cultures by the heavy emphasis on intergenerational ethics. Yanagita saw that people had derived great meaning from achieving something in their lives which could be passed on to the next generation. In the Japanese traditional sense of values, it means that the current generation should not only avoid creating difficulties for the next generation, but ought to feel that one of their reasons for living was to make a better life for them. Here, Yanagita has found a characteristic of Japan: that they think it is their duty and reason for living to preserve nature for the next generation and turn over it to them.

Yanagita gives the following example:

> In the Nasu Mountain area at Hyuga (Miyazaki prefecture, Kyushu), when constructing a suspension bridge with wisteria vines, people select big Japanese cedar trees for the four corners of the gate tower and bind the longest vines to the trees. The cedar trees look to be eighty to ninety years old, although they are still in good condition, but there are good young trees by the side of them on the four corners, which it seems were planted 10 years ago. I thought that when those trees became useful, the village and all its people would have regenerated, and it was deeply embedded in my mind that village life would be everlasting. (*Mame no Ha to Taiyo* [Bean Leaves and the Sun]).

I will talk later about why Yanagita had come to think of the Japanese sense of values and ethics in this way.

5.4   Perspective on Nature, Ethics and Faith

I will now consider the first point, the idea that nature does not exist just for human beings but has inherent value. Yanagita thinks such a perspective is a part of the traditional Japanese view of nature. He sees that this perspective has played a role in limiting the destruction of nature in rapidly industrializing Japan.

What are intergenerational ethics and this perspective on nature based on? Yanagita thinks these ideas rely on "*ujigami* faith". Here, "traditional perspective on nature and ethics" refers not only to premodern times. Yanagita argues that this perspective continued into modern times, the post-war period.

In modern Japan, in each region – urban areas as well as villages – there were small shrines to *ujigami-sama* (tutelary deity) or *ubusuna-sama* (guardian deity of one's birthplace), and *omiya-san* (Shinto shrines) surrounded by woods or forest. Such shrines remain in many regions up to the present. *Ujigami* worship is faith in a deity who was deified by the locals.

It is generally understood that Japanese culture and way of thinking were influenced by Buddhism and Shinto. Shinto is used in a wide sense here to refer to the general Japanese faith in *kami*. Yanagita divided Shinto in the wide sense into two groups. One is a faith typified by large shrines such as the Ise or Atsuta. This is the Shinto which has doctrines based on *Kojiki* (Records of Ancient Matters) or *Nihonshoki* (Chronicles of Japan). It corresponds to the study of Japanese classics by Norinaga Motoori, and the state Shinto of the prewar period. The second Shinto group is the *ujigami* worship of ordinary people – that is the faith in small shrines. Here, Yanagita argues, the fundamental way of understanding or thinking of ordinary people is based on *ujigami* worship rather than on Buddhism or the *Kojiki* and *Nihonshoki*. More than just a tradition, it has been maintained and handed down to post-Meiji modern times.

5.5   *Ujigami* Worship, Ancestors and Descendants

What is the specific content of *ujigami* worship? How have the Japanese perspectives on nature and intergenerational ethics arisen in *ujigami* worship? What is Yanagita's view? He is a folklorist and infers the following from his research.

The *ujigami* was believed to be a fusion of generations of ancestor spirits. When people died, after a certain period the spirit joined the fusion, becoming an *ujigami*.

For the *ujigami* to fuse, memorial services needed to be held by descendants for a specified period.

Therefore, there were three types of soul among the Japanese. One was a spirit who fused with the *ujigami* a certain period after death. Another was a spirit on its way to becoming the first type, who was receiving memorial services from its descendants. The last type was a sprit who received no memorial services from its descendants, and so could not become a deity. This spirit had no descendants to hold memorial services, or whose descendants had died.

Regarding the location of the *ujigami*, rather than dwelling apart from this world like Buddhism or Christianity, they live on the top of mountains near each village and watch over the people who live there. They then visit the village in spring and autumn. At that time, the villagers conduct a ritual: the village festival.

Yanagita writes:

> It is thought that at the end of this life, we are sent into a deep, shady valley, and there separate from all pollution and gradually rise high up into the air. In earlier times people imagined that our ancestors' spirits, already purified, rest on the mountains in bluish clouds and look over the wide country from afar. (*Yamamiya-ko* [The Study of Mountain Shrines]).

The Japanese view of *ujigami* worship was that the *ujigami* look over the life of their descendants. Please do not think that I am discussing this on the assumption that *ujigami* actually exist; this is rather the world of ideas. These ideas exist in the minds of ordinary people.

Even today, many people visit shrines and pray on New Year's Day or when a new baby is born, and on a festival day for children of three- and seven-year-old girls, and three- and five-year-old boys. Yanagita argues that this is a ritual based on *ujigami* worship, although mixed with Buddhist rites introduced from China, and other elements.

According to this idea of *ujigami* worship, children, as well as the aged or hand-icapped, should be respected as they will become deities. Yanagita writes that this is a fundamental base for a sense of ethics to support people's coexistence. However, it was considered that for a deceased person to fuse with the *ujigami*, descendants must hold memorial services for quite a long time. Moreover, the *ujigami* festival should be held by villagers who are the descendants of the deceased (Fig. 5.2).

Therefore, the Japanese take a serious view of the permanent prosperity of their own descendants, in other words the "family longevity". Due to this, the Japanese have strong family-centered values. After death, to fuse with the *ujigami* and become a deity, a spirit needs descendants to hold memorial services. Moreover it is expected that the family will be succeeded by generations and as villagers they will worship *ujigami* in spring and autumn. These are necessary conditions. This provided people with purpose in their lives: maintain the family line, bring up children, worship the ancestors and then after death become a deity and watch over descendants. As I mentioned earlier, for the Japanese, intergenerational ethics or considering the next generation is the important purpose for living, and it derives from this point. It is one of the most important elements of Japanese values.

Fig. 5.2 *Ujigami* worship and intergenerational ethics (festival of *ujigami*). *Source* Photo courtesy of Koichi Kato

On the other hand, it was believed that a spirit who was not offered memorial services by his descendants, because his line was extinguished, would not only fail to fuse with the *ujigami* but would also cause calamities for people. As a result, people were afraid that they would become a spirit with no descendants, haunting this world. Because of this, people also attach importance to the permanent prosperity of their descendants.

5.6 Purpose of Life to Watch over Descendants

The purpose of life is to work for the continuation of descendants and watch over their future. Yanagita thinks that this idea gave birth to the Japanese values and ethics of finding purpose for life in strong intergenerational ethics, and doing something for the next generation.

Yanagita's view is that the background of the idea that generations of ancestors become a deity which watches over descendants, is the earnest desire of people to protect their descendants and help them even after death. He feels that this desire, rather than being a form of narrow egotism such as wishing only benefits for one's own family members and children or descendants, is a positive thing which leads to intergenerational ethics. People's lives are never conducted in isolation; children's growth, for example, is closely related to the circumstances and social relationships in which they live, such as the community, their culture and nature there, as well as their own family and relatives. Accordingly, the society and culture surrounding them needs to be fertile, so that descendants will grow up in good health, have no great distresses and live full lives. Besides, the nature which nurtures them needs to be preserved.

People consider not only their own circumstances, but seriously think of those of their descendants, and this entails consciously recognizing the relationships between the future of descendants and their society, culture and nature. In practice, the traditional way of thinking means there is a high level of self-awareness of this, and Yanagita's work suggests that this intergenerational ethic is a characteristic of the Japanese. He presents the ideas given here as an example of the intergenerational ethics that have been passed down from generation to generation.

5.7 Kami on the Mountain, Inhabiting Nature

My first point concerned people's perspective on nature. I earlier suggested that the *ujigami* exists as a fusion of the ancestors' spirits dwelling at the mountaintop, and looking down on the village. Yanagita thinks that the Japanese view of mountains as sacred arose from this idea. Mountains are hallowed because of the *ujigami* dwelling there. Because the forest or wood is close to the *kami*, people venerate it as a sacred area, and this became an important reason to maintain a wide range of forestry.

Furthermore, I mentioned earlier that the *ujigami* was thought to be a fusion of spirits, which has no concrete form and so cannot be seen. People imagined that the invisible *kami* came down to the village and inhabited a specific natural object: a plant or tree. A well-known example of this is the Ombashira Festival at Suwa. There, huge pillars are built at the four corners of the shrine. The villagers believe that the *kami* come down the *ombashira* (pillars). That is, it is believed that the *ujigami* comes down to the village and dwells in the form of a natural object. The *kami* targets the natural object as a *yorimashi*, an object representative of a divine spirit, or a *yorishiro*, a vessel for a spirit, so the plant or the tree is considered a sacred thing.

It is often said that Japanese religion is animistic. Animism is the idea that a spirit exists within a natural object or dwells in it inherently. However, Yanagita argues that the Japanese notion of the divine is not animistic. Although the Japanese belief appears superficially animistic, the *kami* are not inherent in a natural object, rather inhabiting natural objects on occasion, and people venerate the natural objects in which the *kami* dwelt. Various animals on mountains were close to the *kami* and they were thus venerated as harbingers when the *kami* came down to the villages.

Therefore, the traditional Japanese idea of nature is not animistic, but it could be said that nature is regarded as having sacred qualities, and is not just for the benefit of human beings but has inherent value. This way of understanding has been maintained traditionally.

It is necessary to obtain the materials which people need for their lives from the surrounding nature, but if these are in a sacred area they have some sacred characteristics. Hence Yanagita believes that they are used with discretion. For this reason, it is rare for the Japanese to think in terms of nature being protected, cut off from human society. Rather, nature is put to practical use in everyday life and protected there. This is the notion Yanagita regards as characteristically Japanese. This idea has often been called "living environmentalism" in recent discussions. I mentioned primatology in Japan earlier. I believe you will now understand why the habitats of Japanese and troupes of monkeys overlap.

5.8   Change in Sense of Value with Modernization

However, after the 1970s, *ujigami* worship as an internal faith was lost among the general public. If that is so, how can the intergenerational ethics and notion of nature based on *ujigami* worship contribute to the view of the environment in the future? Yanagita leaves a hint:

> Even a father hopes that his child will live more happily than him... but the mother, more than the father, cannot help but hope that at least her own children will not suffer the same hardships she endured. Even if they can do nothing for themselves, assuming their children will face the same circumstances they have themselves faced, there is still some hope that they can be of use to the next generation. It used to be that this hope derived only from faith in the powers of prayer and conviction, but we now have some more opportunities offered by education. Good development of Mind and Body to withstand life's hardships is not sufficient, they need further to cultivate their own capacity to question, consider well, and once convinced, to put decision into practice: is this possible?... Parents will never abandon this struggle. (*Meiji Taisho Shi Seso Hen* [Meiji Taisho History: Customs and Manners] Trans. Kenji Sasaki).

Yanagita thinks that *ujigami* worship will disappear eventually with modernization. Whether this is a good thing or bad, he thinks that it is inevitable. In that case, it will be impossible to restore the former beliefs. Yanagita argues that a faith cannot be taught or forced from outside. A common argument is that religious education should be taught in school, but Yanagita directly opposes this. The only way that the former perspective on ethics can be handed down from generation to generation is through education, but when Yanagita says 'education', he is not limiting this to only school education. Rather than this, he believes that it must be handed down intentionally to the next generation within the family or community. This is also the intergenerational issue.

5.9 'Cold' Circulating Society in Traditional Village Community

Furthermore, Yanagita describes a characteristic of the traditional Japanese village community as being a 'circulating' society and he urges the re-evaluation the culture of life from the level of lived experience of modern people caught in the turmoil of modernization. So what message can we understand from Yanagita's argument?

Here a circulating society means not only the society which circulates resources, as the term is often understood. It denotes a "cold society" as described by Lévi-Strauss. A cold society is the society in which the basic structure remains unchanged, even when changes occur in individuals as components of the society. The society as a whole can absorb the changes, and circulates in a homeostatic mechanism. In contrast, Lévi-Strauss calls modern society a "hot society". That is, the whole social system steadily changes as changes occur in individual elements. Lévi-Strauss considers the possibility of cold and hot societies acting together, but he comes to no clear conclusion.

Yanagita also considers, in his own way, how to connect the traditional society and modernization as a kind of circulating society. He published extensive arguments about this issue in the 1920s, writing that modernization is an inevitability. It cannot be avoided, as modernization is industrialization. This kind of technological progress is unavoidable and moreover it is necessary.

5.10 Problems of a Resource Consumption-Based Life

However, Yanagita issues a warning about which direction people are moving to satisfy their spontaneous desires; in other words, towards a lifestyle of resource consumption. He points out that modernization is necessary, but the resource consumption lifestyle may cause serious problems in Japan's future.

What that means is, at that time, East Asia was enmeshed in a rather unstable international situation. In that situation, if lifestyles became based on desire fulfillment or resource consumption, more markets and resources would need to be secured. This would be unsustainable, given the high probability of conflict with China – the major source of resources and markets – or Western powers such as the USA and the UK which wanted a share of China resources and markets. In short, he warns of the likelihood of taking the first steps towards war.

Taking this path leads to ruin. Thus, what Yanagita suggested is reconsidering people's living culture from the fundamental level: not what they may or may not consume, but what they live for. That is, people need to reflect on their own way of living in the midst of modernization and create their own way of life themselves. It is difficult to maintain a circulating society without change as in former times. However, if we adopt the Western style of modernization with its resource-intensive lifestyle, ruin lies ahead. What should we do? For that, we should consider once more, not only about our everyday lifestyle, but also what we live for, the meaning of life and our reason for living.

5.11   What Is a Prosperous Life?

A prosperous life is important, of course. However, a prosperous life does not necessarily involve consuming a lot of things. In order to live, a human being needs what are called daily living necessities and a certain amount of luxury goods. Relating to this, problems had arisen in the situation of farmers and workers in the 1920s. They should have become more prosperous. However if people move towards satisfying their natural desires, consuming their wealth, do they really have more fulfilling lives? Yanagita suggests that instead it is necessary to adopt a point of view which relativizes the value of consumption.

Yanagita does not dictate how people should live; he says that one should consider for oneself what constitutes a fulfilled life and what kind of life might be fulfilling for one's descendants, and thereby create a new culture with one's own capabilities. Human beings need to interrogate themselves as to the purpose of their lives.

Yanagita approached these issues in the turbulent situation of 1920s Japan, yet while listening to the presentations here by eminent scholars, it seems to me that his ideas now have global relevance. The same approach applies to issues of regional environmental protection and the coexistence of human beings with nature.

At the beginning I said that I would like to talk about the problems of environmental pollution in Japan. The traditional Japanese perspective on nature has not only good aspects but also some problems. I will address these at the third session.

References

Kawada M (1997) *Kunio Yanagita – His Life and Thoughts* (in Japanese). Yoshikawa Kobunkan, Tokyo

Kawada M (1998) *Japan as Viewed by Kunio Yanagita – Folkloristics and Design of Society* (in Japanese). Miraisha, Tokyo

Kawada M (2016) *Kunio Yanagita – An Overview of Knowledge and Social Planning* (in Japanese). Chikuma Shinsho, Tokyo

Minoru Kawada Professor Emeritus, Graduate School of Environmental Studies, Nagoya University, Doctor of Law, Research interests: History of political thoughts, and history of politics and diplomacy.

Born in 1947, Professor Kawada completed a doctoral course at the Graduate School of Law, Nagoya University and started to teach at the University's School of Law. He later lectured at

Nihon Fukushi University before taking up a professorship there. He became a professor at the School of Informatics and Sciences, Nagoya University in 1996, and a professor at the university's Graduate School of Environmental Studies in 2001. His publications include: *Study on the History of Kunio Yanagita's Thoughts* (in Japanese), Miraisha, 1985; *Kunio Yanagita – World of Native Faiths* (in Japanese), Miraisha, 1992; *Kunio Yanagita – His Life and Thoughts* (in Japanese) Yoshikawa Kobunkan, 1997; *Japan as Viewed by Kunio Yanagita – Folkloristics and Design of Society* (in Japanese), Miraisha, 1998; and *Osachi Hamaguchi and Tetsuzan Nagata* (in Japanese), Kodansha, 2009.

Part II
International Conflict Concerning Environmental Damage and Its Causes

Chapter 6
Kosa (Asian Dust Particles) and Air Pollution in Asia

Yasunobu Iwasaka

Taking the fine yellow dust particles we call *kosa* as subject matter, I would like to discuss environmental problems based on the idea of understanding the problems across national borders, making a framework for solutions, and using the framework to solve the problems. The 21st century is now often referred to as the age of environmentalism, and people in various fields have argued that Asia will play a central role in determining the trends of that age. The reason I will be discussing *kosa* is not unconnected to this.

In this lecture, I would like to focus on three topics. First, I will briefly review the history of the study of *kosa*. This does not mean simply considering the research in chronological order. Rather I will show at the same time how with regard to cross-border phenomena, the government and researchers in different countries changed their attitudes, and gradually deepened their understanding. Second, I will deal with the topic of the relation between Earth's environmental changes and *kosa*. You will understand this as I continue my lecture. This viewpoint has played an important role in building the current cooperative system among Japan, China, Korea and Australia. Thirdly, I will discuss the present situation of *kosa* research and environmental conservation in this area.

6.1 The Study of *Kosa* in Japan

Figure 6.1 is a summary of how *kosa* research has been conducted in Japan. *Kosa* research started in earnest in fields such as meteorology and atmospheric science in the 1960s and 1970s. In this period the research was mainly concerned with whether *kosa* would become ice nuclei or condensation nuclei. Such research started as a result of energy problems Japan was facing at that time.

Yasunobu Iwasaka, Professor Emeritus, Graduate School of Environmental Studies, Nagoya University; former Professor, Frontier Science Organization, Kanazawa University.

© The Author(s), under exclusive licence to Springer Nature Switzerland AG 2020
Y. Hayashi et al. (eds.), *Balancing Nature and Civilization—Alternative Sustainability Perspectives from Philosophy to Practice*, SpringerBriefs in Environment, Security, Development and Peace 32, https://doi.org/10.1007/978-3-030-39059-4_6

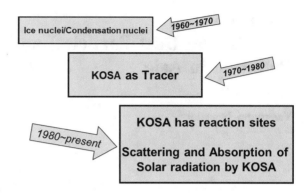

Fig. 6.1 History of *kosa* research in Japan. *Source* The author

Almost all energy at that time depended on hydraulic power generation. The foundation of Japan's energy supply was held to be securing or building high quality water sources. In those days, meteorologists were researching whether it was possible to cause rain. If that had been possible, it would have been easy to secure water resources. The term "artificial stimulation of rain (artificial rain)" frequently appeared in the newspapers at that time. To cause rain, it was necessary to produce clouds, and ice nuclei were required to make clouds. *Kosa* research of that period seems to have attracted a lot of attention worldwide, and some textbooks in foreign countries introduced *kosa* research as something like "peculiarly Japanese research". When I read these sections, it seemed to me that the implication was that unique research had been achieved with extremely small sums of money. The idea of planning a grand project of rainmaking, using easily available sand is quite fascinating, isn't it? In any case, the energy problem was the driving force of *kosa* research in those days.

Time passed, and the main source of energy dramatically shifted from hydraulic power to thermal power. Everywhere in Japan, problems like public pollution surfaced in the 1970s and 1980s. Initiatives to expand industry slowly but surely emerged, not only in Japan but in the whole of Asia. Some researchers started to feel apprehensive about the possibility of pollution produced by industrial activity in China with its population several times larger than Japan's. Of course, Japanese industries had been emitting pollutants. After these became serious social problems, businesses were required to take various countermeasures. It was also pointed out that some pollutants were carried over from the developing country China.

6.2 *Kosa* as a Tracer, *Kosa* Reacts

The strongest westerly winds blow in the Far East. Nobody doubts that various particles are being blown from the Asian continent to Japan and the Pacific. The issue of pollutants crossing borders was discussed by those with an interest in the matter, but in those days, measuring technologies were still underdeveloped. Of the airborne particles, at that time *kosa* was relatively easy to detect. The idea of using *kosa* as a tracer therefore arose: if *kosa* was coming from China, pollutants might also be coming in the same way. This idea ushered in an age in which attention turned to the long-range transport of *kosa*.

Subsequently, from the middle of the 1980s up to the present, *kosa* has been considered to play a very important role in Earth's environment. One of the concepts underlying this viewpoint is that *kosa* is not simply flying. While *kosa* is airborne, it is reacting chemically with other nearby airborne particles. Furthermore, *kosa* may be blocking, absorbing or reflecting radiation from the sun. These concepts have been increasingly regarded as factors that cannot be neglected when forecasting the future of Earth's environment.

6.3   Observation and Tracing of *Kosa*

The above-mentioned period when *kosa* came to be regarded as a tracer gave us an opportunity to reconsider the relationship between the environment and nations. Proving scientifically that *kosa* flies quite extended distances, or that *kosa* observed in Japan actually originates in China, are very difficult tasks. Suppose our meteorological agency in Japan announces: "There will be a lot of *kosa* today." If we collect air during this period, we will catch a lot of sand. However, we would not be able to assert that the sand collected was *kosa*. The sand would not be that much different chemically or mineralogically from the sand we see elsewhere. Basically it is the same. Supposing the sand was labeled "Made in China", it would be very easy to ascertain its origin. However, since it is not actually labeled, it is extremely difficult.

The period from the 1970s to the 1980s was a time when various electronic and communication technologies made remarkable progress, and remote-sensing systems called Lidar, and satellites (such as the Japanese Himawari weather satellites) were launched. The time had come when we could observe *kosa* with such systems. In the Lidar system, a laser beam is projected into the sky, and the system clock starts timing. The system measures the time until the laser beam returns after hitting something. From this time, the system calculates how distant the reflecting objects are. The development of such systems now enables us to answer to some extent such questions as how *kosa* flies over to us, and whether *kosa* really does fly over to us.

6.4   *Kosa* Observed by Lidar

Figure 6.2 is the first report of a *kosa* observation made by Lidar not only in Japan but in the world. The vertical axis represents altitude, and the horizontal axis represents backscatter coefficient, which is the technical term for *kosa* density. The results obtained by Lidar observation show that high *kosa* density is found at two

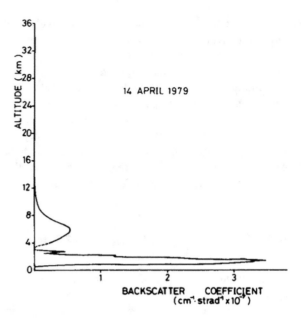

Fig. 6.2 *Kosa* observed with Lidar at Nagoya (April 1979). *Source* Iwasaka et al. (1983: 191) CC BY 4.0

altitudes: 2 and 6 km. This information was very meaningful. In those days, we did not have clear evidence to show how high the *kosa* flew. It was vaguely assumed that *kosa* was flying near ground level, but actual observation showed that high density areas existed at much higher elevations above the ground, and the sand then precipitated: this was the *kosa* phenomenon. In this way, we now can estimate the location of high *kosa* density areas, and from where *kosa* comes based on accurate observation.

Figure 6.3 shows the path of an air mass which arrived above Nagoya at noon on April 14, 1979. Its path can be traced back over the course of previous days. As expected, the result proved that the air mass containing a lot of *kosa* came from China.

Interested parties had imagined that pollutants also came to Japan in the same way. However, in the 1980s it was taboo to express that idea. When I introduced the results of this observation to a research group, the professor heading the group told me clearly never to say at an international conference that chemicals such as sulfur oxides might also be coming in the same way. That was the situation in the 1980s. The professor had a lot of friends in China and was known as a China expert among researchers. When I was warned in this way, I used to reply that I was only sharing the results in that group; I would never report them in my thesis. From this verbal exchange, you can get an unexpected glimpse of how the attitudes of researchers reflect the relations of their country to another country, and national policy.

So, we had an understanding to some extent that some airborne pollutants were traveling from the Continent to Japan and the Pacific. But it was quite difficult to discuss the matter further. These problems were still unsolved when another serious issue appeared on the stage.

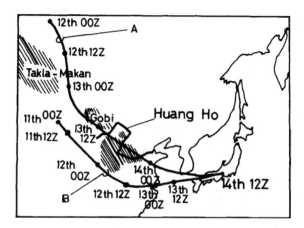

Fig. 6.3 Air masses from China observed with Lidar at Nagoya (April 1979). Curve A was computed for the potential temperature θ = 316.4 K and curve B for θ = 292.2 K. *Source* Iwasaka et al. (1983: 193) CC BY 4.0

6.5   Environmental Change and *Kosa*

Quite a few people may know Fig. 6.4 which appeared in the 2001 IPCC report (IPPC stands for International Panel on Climate Change). It lists the major agents that are related to global warming or control of the temperature of the earth. In the

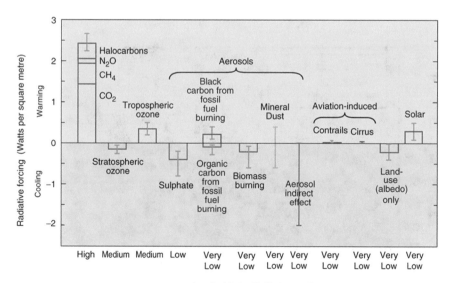

Fig. 6.4 The IPCC third report. The global mean radiative forcing of the climate system for the year 2000, relative to 1750. The term "Mineral Dust" appears in the middle as a component of Aerosols. *Source* IPCC (2001: 8)

vertical direction, you will see how these agents work. Using the vertical axis '0' as the border, if the bar goes upward, the agent warms the earth compared with the year 1750; if the bar goes downward, the agent cools the earth. The length of the bar shows the warming/cooling rate.

This 2001 IPPC report is the third version, and the term "mineral dust" appeared for the first time in this version. This is a familiar substance known to us as *kosa*. The bar extends both to the warming side and to the cooling side. This means that it is undecided whether *kosa* works to warm or cool the earth. Under the graph, the degree of scientific understanding is summarized with the expressions 'high', 'low', etc. For example, CO_2, Methane, N_2O work to warm the earth, since the bar goes higher than the 0-axis, and the degree of understanding is fairly high since it is described as 'high' at the bottom. When it comes to mineral dust, however, the effect is not yet known, or further study is needed since the degree of scientific understanding is described as "very low". From around the time this report was issued, the world of *kosa* study started to change drastically, as did the attitudes of governments.

Figure 6.5 shows the trend of CO_2 density, which is one of the substances considered as having "high degree of scientific understanding". This is very famous observation data obtained at Ryori, in Tohoku, Japan (Ryori, Sanriku-cho, Ofunato-city, Iwate Pref.), an observatory at the top of Mt. Mauna Loa, Hawaii, and from the south pole. The data at each point shows the annual increase. Scientific explanations have been given for the differences in data values, and the understanding of CO_2 has progressed fairly well. Although there remain some points that we do not understand very well, we are now at a stage where various measures have to be taken.

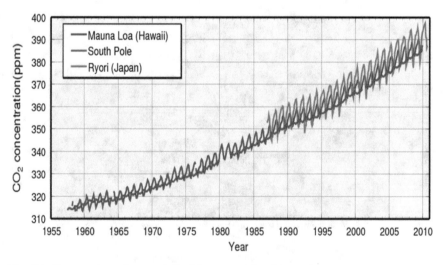

Fig. 6.5 Concentration of atmospheric CO_2 measured at Mauna Loa, Ryori and South Pole. *Source* The Japan Meteorological Agency (2004: 31)

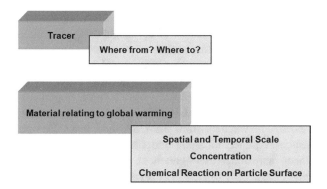

Fig. 6.6 Change in standing point of *kosa. Source* The author

As for *kosa*, we do not know it at all, but it has been decided that *kosa* should be taken up as a subject for research. This is such a big change from the time *kosa* was dealt with as a tracer (Fig. 6.6). When *kosa* was dealt with as a tracer, in a sense our job ended as long as we found where the *kosa* came from and where it went. However, from the time when *kosa* was taken up as an agent related to global warming, it became necessary to pursue the quantitative aspects of *kosa:* what is its density? For how long and where is it airborne? In particular, while *kosa* is airborne, let us suppose it 'eats' SO_2 etc. This expression may sound strange, but suppose *kosa* attracts SO_2 to the surface. There is a possibility that this may cause a serious problem, which is to drastically change the idea that sulfur mist cools down the earth. (In the global warming forecast, we usually suppose that emitted SO_2 will work as sulfur mist, but such a major supposition will not hold true in the above process.) It is necessary to clarify the chemical function of *kosa*. At any rate, researchers are increasingly being asked to supply quantitative data or chemical understanding, while governments are now being asked to handle *kosa* as an agent of global warming rather than as a tracer. Thus, *kosa* research had the effect of making researchers and governments change their ideas and attitudes.

6.6 A New Stage: Towards Cooperative International Research

These new demands made acceptable what had previously been taboo. Research was moving to a new stage supported by new demands. American and especially Japanese researchers thought that there was very little information available in the area where *kosa* originates. For the sandstorm phenomenon that we call *kosa*, it is a rule all over the world to put an 'S' on the weather map. Since this rule is commonly followed in China, Korea, and Japan, Japanese researchers took it for granted that other researchers had the same understanding, but we started to realize that it was a big mistake to think this way. In this sense, Japanese researchers had not been attentive to the circumstances of the neighboring countries (for example, observation methods, legal standards etc.).

On the other hand, observations by international teams started to progress in earnest in Japan, America, China, Korea and elsewhere. After the shocking results published in the IPPC report, many projects were formed. In particular, researchers from many countries joined in the ACE-Asia project. This project began under the leadership of NASA researchers from the USA, and many European as well as American researchers joined it. Japan and China had made a science and technology agreement, and under this agreement, Japan has been taking part in the project, evaluating the supply of Aeolian dust and its influence on the climate. The English project name for this is "Aeolian Dust Experiment on Climate Impact", which has a slightly different nuance from the Japanese. I was personally involved in this project, and Prof. Kai, my colleague, played an important role.

6.7 Collection of *Kosa* at Its Origin

One of the sources of *kosa* is the Taklamakan Desert. We became aware of this desert, and decided to conduct a field study. The place we actually made our camp was Dunhuang, to the east of the desert (Fig. 6.7). Since the wind generally blows from the west, it comes from the center of the Taklamakan Desert towards Dunhuang. Thus, if we wait there, the wind over the desert will come to our camp

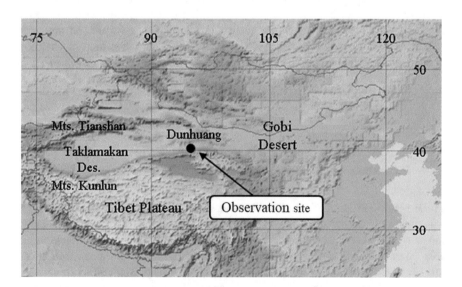

Fig. 6.7 Balloon-borne observation and ground-based Lidar observation at Dunhuang with Japan/China cooperation. *Source* The author

Fig. 6.8 Balloon-borne particle sampler. *Source* The author

site. Dunhuang is also a sightseeing spot, and that was another reason we chose it as our site for observation.

Figure 6.8 shows a scene from the *kosa* observation. We flew a balloon with a *kosa* sampler (collecting machine). The sampler was suspended from a balloon with a parachute between the balloon and the sampler. The sampler was protected by egg boxes. We did not buy eggs there, but just bought egg boxes, and used them as cushioning material. The inside of the sampler was made in Japan, and the outer material was Chinese. It was, so to speak, a jointly-made Sino-Japanese measuring instrument.

Figure 6.9 shows the internal structure of the sampler on the left. There are air intake apertures, and *kosa* is collected through the apertures by taking in air. The construction of the entire balloon is shown on the right. Between the balloon and the parachute there is a cutter. Where the sampler moves is checked using the attached GPS (Global Positioning System). When sampling is finished, the cutter is instructed to cut the rope, and the parachute and lower part then drop to the ground, where we recover them, using a Jeep. When we used this device, we asked the Chinese researchers to obtain the appropriate radio transmission license, as well as other necessary legal permissions. Conducting a project jointly between Japan and China or another country gradually brings us to some sort of mutual understanding.

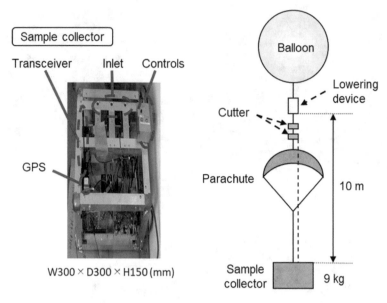

Fig. 6.9 Particle sampler mounted on balloon. *Source* The author

6.8 *Kosa* Collection Sites and Chemical Components

Now, I will turn to something a little technical. Let us examine the pollutants attached to the surface of *kosa*. When we examine the samples collected in the upper atmosphere, we can see variously shaped minerals. We can also determine the chemical composition of these minerals. For example, typical minerals contained in the sand are Si (silicon), Ca (calcium), Al (aluminum), SiO_2 (silica), and Na (sodium). By analyzing each grain of sand to find the density of aluminum and the density of calcium, we can obtain triangular diagrams as shown in Fig. 6.10.

Diagram (f) in the bottom right is for the samples collected over Japan. The other diagrams are for the samples collected in Dunhuang. The samples of the top three diagrams were collected over Dunhuang at altitudes of 5–7 km and 3–5 km, and in different seasons. As mentioned earlier, aluminum and calcium are typically contained in the minerals. Thus, if the sample is something like a mineral, the particles come closer to the right side of the triangle. The samples collected on the ground at Dunhuang and the samples collected over Dunhuang show this tendency.

'S' is sulfur. Some particles contain a lot of sulfur; for example, the samples over Japan. The *kosa* samples collected over Japan have many particles shifted to 'S'. After further research, we thought it was probably because the airborne *kosa* became a sulfuric compound by absorbing sulfur dioxide gas. This does not tell us where such a process occurred. However, at the original site, the *kosa* was clean, yet by the time the *kosa* sand arrived over Japan, it was polluted. Some people said that it was polluted in Japan, but that was unthinkable as the sampling was carried out

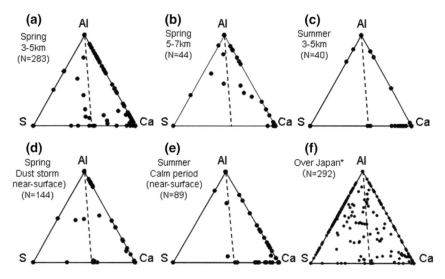

Fig. 6.10 Relative weight ratios of Al, S and Ca for the mineral particles collected in Dunhuang (**a–e**) and in the free troposphere over Japan (**f**). *Source* Trochkine et al. (2003: 8). Reproduced and modified by permission of American Geophysical Union. © American Geophysical Union, 2003

over Wakasa Bay, about 5 km above sea level. Thus, *kosa* encountered sulfuric dioxide gas while crossing China en route to Japan. The project to examine the origin of *kosa* unexpectedly raised the problem of detecting early signs of pollutants. Of course, this report was published in collaboration with the Chinese professors. Nowadays, as it is generally understood among researchers that there is a strong possibility that pollution on the surface of *kosa* is caused by the polluted air emitted from the industrial areas in China, it is possible to issue a joint report.

6.9   Building Cooperation Among Asian Countries

What are governments doing? At a meeting of the environmental ministers of Japan, China, Korea and Mongolia several years ago, it was agreed to handle *kosa* as a shared environmental problem. In this move to tackle *kosa* as an environmental problem, discussions are ongoing to raise common problems for each country and identify what cooperation and assistance are mutually needed. Some programs have been carried out already. In Japan, a *kosa* research committee (the Special Committee on Dust and Sandstorm Issues) was set up in the Ministry of the Environment, and I am acting as chairman. We are now working on measures for Japan to take based on the most up-to-date scientific knowledge. Other countries are doing the same.

Now, the atmosphere of our discussion has changed to the extent that it would not be so tense even if we declared that *kosa* flies over borders, so pollutants may be doing the same. Regarding the definition of *kosa*, we now have a mutual and deeper understanding about the fact that each country has different definitions. Definitions reflect the culture of each country, and the attitudes of its people towards nature, so we are now of the opinion that we do not follow one particular method, but we start from an understanding that we have different methods of our own. I hope the time will come in the near future when we can sit at a table to talk frankly about the problem of various pollutants crossing borders as well as the problem of *kosa*. Our handling of the border-crossing problem may be far behind the situation in Europe, however, that is not necessarily bad. Considering the diversified development of Asian cultures and the length of history, we can't make progress overnight. Through our research, scholars have come to a common understanding, and have published jointly. I am glad to report that governments are also to some extent in the mood to discuss freely based on the common scientific knowledge.

References

Intergovernmental Panel on Climate Change (2001) *Climate Change 2001: The Scientific Basis. Contribution of Working Group I to the Third Assessment Report of the Intergovernmental Panel on Climate Change [Houghton JT, Ding Y, Griggs DJ, Noguer M, van der Linden PJ, Dai X, Maskell K, Johnson CA (eds.)].* Cambridge University Press, Cambridge and New York

Iwasaka Y, Minoura H, Nagaya K (1983) The transport and special scale of Asian duststorm cloud; a case study of the dust storm event of April 1979, *Tellus B* 35: 189–196

The Japan Meteorological Agency (2004) *Climate Change Monitoring Report 2003*

Trochkine D, Iwasaka Y, Matsuki A, Yamada M, Kim Y.-S, Nagatani T, Zhang D, Shi G.-Y, Shen Z, (2003) Mineral aerosol particles collected in Dunhuang. China and their comparison with chemically modified particle collected over Japan. *J. Geophys. Res.*, 108(D23):8642. https://doi.org/10.1029/2002jd003268, ACE10 1–11

Yasunobu Iwasaka Professor Emeritus, Graduate School of Environmental Studies, Nagoya University; former Professor, Frontier Science Organization, Kanazawa University, Doctor of Science, Research interests: atmospheric aerosol, atmospheric physics.

Born in Toyama Prefecture in 1941, Professor Iwasaka graduated from the Faculty of Science, University of Tokyo in 1965 and completed a doctoral course at the university's Graduate School of Science in 1971. After working at the Water Research Institute, Nagoya University, he became a professor at the university's Solar-Terrestrial Environment Laboratory in 1989, and at the Graduate School of Environmental Studies in 2001. He became a professor at the Institute of Nature and Environmental Technology, Kanazawa University in 2004 and is currently a professor at the university's Frontier Science Organization. He was awarded a prize by the Meteorological Society of Japan in 1990. His publications include: *Ozone Hole – Earth's Atmospheric Environment Observed from the Antarctic* (in Japanese), Shokabo, 1990; *Introduction to Environmental Sciences 2 – Atmospheric Environmental Science* (in Japanese), Iwanami Shoten, 2003; *Asian Dust – Looking into its Mysteries* (in Japanese), Kinokuniya, 2006; and (co-editor) *Yellow Sand* (in Japanese), Kokonshoin, 2009.

Chapter 7
Environmental Charges Levied on Heavy Goods Vehicles in the EU

Werner Rothengatter

The European Union is not a homogenous decision-making unit. The Union has a limited legal competence while the single member states still maintain highly autonomous decision-making, including in the transport sector. This will be briefly summarised in Sect. 7.1. From this it follows that there is also a considerable heterogeneity of charging systems for heavy goods vehicles (HGVs) as presented in Sect. 7.2. Section 7.3 includes a comparison of three HGV charging systems for the EU countries, Austria and Germany, as well as for the non-EU country Switzerland, which is surrounded by EU countries and well-known because of its rigorous policy towards road transport to protect the environment. Section 7.4 presents the development of EURO categories as a base of environmental charges and the progress in developing an interoperable payment device to overcome the problems of different charging technologies in the EU. Section 7.5 concludes by discussing the impacts of the different charging schemes on the freight transport market, taking the total revenues as a proxy for the cost burdens which the road hauliers have to bear and the forwarders have to calculate when making their modal decisions.

7.1 Decision Competence for the EU Transport Sector

The European Union consists of 27 member states (after the Brexit), 6 of them founder members as of 1957 (Treaty of Rome). The EU was extended to 15 until 1995, to 25 in 2004, to 27 in 2007 and 28 in 2013. In the United Kingdom a referendum was held in June 2016 which came out with a majority of 51.9% of votes to leave the EU. This 'Brexit' is being prepared for 2020 while its conditions are still open. The distribution of legal decision power in the EU is governed by the principle of subsidiarity, which means that the member states should only allocate

Werner Rothengatter, Professor Emeritus, Karlsruhe Institute of Technology, Germany. Email: werner.rothengatter@kit.edu.

© The Author(s), under exclusive licence to Springer Nature Switzerland AG 2020
Y. Hayashi et al. (eds.), *Balancing Nature and Civilization—Alternative Sustainability Perspectives from Philosophy to Practice*, SpringerBriefs in Environment, Security, Development and Peace 32, https://doi.org/10.1007/978-3-030-39059-4_7

decisions to the EU if the associated problems cannot be solved satisfactorily on the national level. For example, the EU is not competent for foreign policy and defense, and in the case of fiscal decisions a unanimous vote is necessary. For other so-called supra-national areas, a majority vote is required, in particular for the areas Single European Market, Environmental Policy, Health or Consumer Protection. The transport sector is part of the Single European Market such that regulations are decided on the European level. This does not hold for transportation investments for which the member states are competent because of their fiscal autonomy. This is the reason why the European policy makers try to motivate the member countries by substantial co-financing activity to follow European plans for establishing "Trans-European Networks". The latter are intended to contribute to environmental goals set at the European level and foster a modal shift to more environmentally friendly transportation modes like railways, inland waterways or coastal shipping.

The political bodies of the European Union are the European Commission (EC), the European Council and the European Parliament (EP). The European legacy consists of Directives which are prepared by the EC and decided by the Council and the EP. The Council consists of all Ministers of member states which are responsible for the area of decision making. A Council of Prime Ministers is called the "European Summit" and decides on basic problems of supra-national EU policy. The EP consists of members elected by European elections and decides together with the Council on Directives prepared by the EC. This means for the case of transport, that Directives have been decided at the EU level on all basic aspects of regulation and pricing for the transport sector.

The EU legacy on HGV tolling has started with Directive 1999/62/EG which sets a framework for the introduction of HGV tolls for all member states which plan to establish such systems. This means that the countries which had already introduced concession systems, such as France, Italy or Spain, were not obliged to change their systems. The network to be charged was restricted to motorways and freeways with a similar standard and dedicated rules were set for the calculation of infrastructure costs, which served as the basis for toll setting. Although the EC had published several green and white papers on fair and efficient pricing in the transport sector, which suggested marginal cost pricing, the Directive made a cost calculation method compulsory which is based on the allocated average costs of trucking. This was influenced by the Council under the pressure of all member states which were interested in the first instance in a full infrastructure cost recovery. The Directive allowed for a differentiation of tolls on the base of the environmental performance which was documented by the EURO standards (EURO 0 was the worst and EURO V the best (forthcoming) standard in 2005[1]). The sum of revenues after differentiation should not exceed the total allocated infrastructure costs of HGVs.

The above Directive was the basis for the introduction of HGV tolling schemes on motorways in Austria (2004) and Germany (2005). It was extended in the year 2006.

---

[1]EURO VI became obligatory for newly licensed trucks and buses in 2013.

Additional roads of the Trans-European Road Networks could be included; the range of differentiations was extended and additional criteria for toll differentiation on the base of congestion could be considered. But the breakthrough for the charging of environmental costs came in 2011, when Directive 2011/76/EU was decided. This Directive is still in place (2020) and includes the following essentials:

- Tolls levied according to distance travelled, type of vehicles and emission category[2];
- All HGVs exceeding 3.5 tonnes gross weight on motorways and other road networks, to be defined by member states;
- Mark-ups are possible for mountainous areas (up to 25%);
- Charges oriented at average infrastructure costs, differentiated by vehicle characteristics (axles, weight), EURO categories and optionally by congestion;
- An additional external cost charge can optionally be added (for air pollution and noise);
- Revenues should preferably be used for the Trans-European Transport Network.

This implies that a charge for external costs can be added in addition to the infrastructure costs, which means that environmental aspects can be included in two ways: First with the differentiation of infrastructure cost-related charges and secondly with the extra charge for external costs. While this appears to be big progress with an environmentally-oriented charging policy, the actual impact was modest, for the following reasons:

- Member states are free to include the external costs in their tolling systems;
- Only external costs of air pollution and noise can be included;
- The regulations for the calculation of the external charges allow only for relatively modest magnitudes ("upper caps") while the requirements for introducing noise charges are very restrictive;
- The option of introducing external cost charges coincides with the phase of very low interest rates on capital markets after the economic crisis in 2008 and the following years. As interest on capital employed makes a high percentage of the capital costs of infrastructure the infrastructure-related costs decreased by a larger amount than external costs could be added such that the sum of HGV charges could decrease (as happened in Germany).

---

[2]The Directive allows member states to continue with applying vignette systems. Belgium, Denmark, Luxembourg, Netherlands and Sweden had introduced a "Eurovignette system", together with Germany and Austria. The latter countries left this time-related system in 2004/2005 and changed to a distance dependent system which is the focus of this paper.

7.2   Heterogeneity of HGV Charging Systems in the EU

Directive 1999/62/EG and its revisions are only relevant for member countries which have introduced new HGV charging systems after 1999 or are planning to do so. All charging systems which existed before, as for instance the concession systems in France, Italy or Spain or the vignette systems in Benelux and Scandinavian countries are not affected by this regulation. Furthermore, countries like Norway and Switzerland, which are close neighbors, have developed their own systems. Figure 7.1 illustrates this heterogeneity graphically.

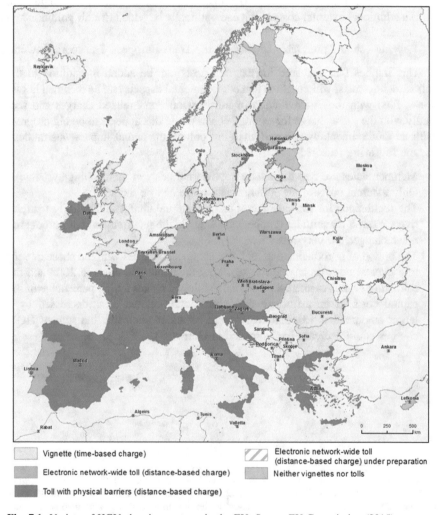

Fig. 7.1   Variety of HGV charging systems in the EU. *Source* EU Commission (2015)

- Distance-related electronic tolling systems based on GPS or Galileo satellite communication are only applied in Germany, Hungary and Slovakia.
- Dedicated short range communication systems are applied in Austria, the Czech Republic, Poland and Portugal.
- Toll plazas are used for manual or electronic payments in France, Italy and Spain.
- Manual toll plazas exist in Greece, Slovenia, Croatia and Ireland.
- Vignette systems exist in the UK, Sweden, Denmark, the Benelux countries, Lithuania, Latvia, Bulgaria and Romania.
- Finland and Estonia do not apply road tolling for HGVs at all.

Only Austria, Germany and Switzerland – as a non-EU country – apply dedicated differentiations and mark-ups to internalize external costs of HGV transport at least partially. The applied technical systems and economic/environmental foundations of charging are presented in the following section.

7.3   HGV Tolling in Switzerland, Austria and Germany

7.3.1   Switzerland

I start by presenting the Swiss case because it represents the most stringent application of the "user/polluter pays" principle in Europe.

(a) *Technical system*

The Swiss HGV charging technology is distance-based and consists of three basic components:

- The distance travelled on all Swiss roads is charged.
- The charging device ('Tripon') is linked to the odometer of the vehicle and is given free of charge to the road hauliers.
- The type of vehicle (max. weight and environmental category) has to be entered into the charging device and is controlled to avoid manipulation.

As the charges don't vary according to the category of roads used the technology is simple and economical. The only infrastructure-related sensors needed are installed at borders to switch off the system when trucks are leaving or to switch on the system when trucks are entering the country. In addition to this general network-related tolling some Alpine passes are tolled separately.

(b) *Economic/environmental foundation of charges*

The charging system for HGVs with a gross weight of more than 3.5 tonnes was introduced in the year 2001 and is based on distance, (weight dependent) average cost of infrastructure and emission category, indicating external costs. As Switzerland changed in the year 2005 to the EU weight limit of 40 tonnes for

Table 7.1 Charges for HGVs in Switzerland since 2017

HGV category	EURO category	Tariff (EU cts./tkm)
I	0, I, II, III	2.77
II	IV, V, EEV	2.35
III	VI	2.01

Source The author. See also Swiss Federal Customs Administration (2017)
tkm: Rate per ton and per kilometre

HGVs (from 28/34 tonnes before 2001/2004), the country tried to apply high charges to prevent the freight transport, in particular the transport crossing the Alps, from diverting to the road. In 2005 the charge was set to 1.45 EU cts./tkm. The tariffs have up to the present increased to the levels exhibited in Table 7.1.

For the frequently used routes between Italy and Germany through Switzerland (about 300 km) a 40-tonne truck with EURO VI category has to pay 240 EUR. This is a multiple of the HGV charges levied in Austria and Germany and has achieved the result that trans-Alpine HGV traffic through Switzerland has not increased substantially. However, not all HGVs have been diverted to the rail mode; a considerable share has been diverted to other routes avoiding Swiss roads, e.g. through Austria.

Two thirds of the revenues from HGV charging are allocated to the Swiss federal government which uses them to finance railway investments, in particular the so-called NEAT tunnels (cross Alpine railway tunnels) through Switzerland. The Gotthard base tunnel is the longest railway tunnel of the world and was opened in December of 2016. Like the other major railway projects, it was financed by an infrastructure fund which is fed by tax hypothecations, allocations from the federal budget and revenues from user charges, including HGVs on roads. The global goal (no growth of trans-Alpine road freight traffic in Switzerland), the associated investment plans (NEAT tunnels) and the financial scheme have been accepted by the Swiss population by referendum.

7.3.2 Austria

Austria has established a state-owned private company, the ASFINAG, which plans, constructs, finances and operates the Austrian motor- and freeways. ASFINAG receives the revenues from road tolling, as there are revenues from vignettes (since 1997) and from distance-based HGV charging (since 2004).

(a) *Technical system*

The HGV charging system applies to all vehicles exceeding a gross weight of 3.5 tonnes. The communication system applied is the Dedicated Short Range Communication System (DSRC). At every entry/exit gantries are installed which

Table 7.2 HGV tolls in Austria for 2019[a]

Euro category	2 axles (EUR/vkm[b])	3 axles (EUR/vkm)	4 + axles (EUR/vkm)
0–III	0.2287	0.3208	0.4620
IV	0.2087	0.2928	0.4288
V, EEV	0.2024	0.2840	0.4188
VI	0.1882	0.2640	0.3944

Source The author. See also ASFINAG (2019)

[a]Day tariffs. Night tariffs are only marginally higher because the mark-ups for noise are almost negligible

[b]vkm: vehicle-kilometre

send signals to the on-board units (called "go boxes") to start or terminate the counting of kilometers travelled. The on-board units are simple to use and the payment can be organized by pre- or post-payment. The installation of the gantries and sensors at the roads can be costly if there are many entries/exits in agglomerated areas.

(b) *Economic/environmental foundation of charges*

The foundation of charges follows the EU Directive and consists of vehicle and environmental characteristics. These are the number of axles, which indicate the deterioration of the infrastructure and space needed, and the EURO categories (see Table 7.2).

In addition to the general system of HGV tolls there are a number of special road sections, including bridges, tunnels and passes, for which special tariffs apply. All revenues go to ASFINAG, which finances operation, maintenance, replacement and new investments from this income. ASFINAG is obliged to pay back all debt acquired for the road system until the year 2047.

7.3.3 Germany

The German system of motorway toll for HGV was planned to be introduced in fall 2003. The provider of the technology was Toll collect, a private company established by Daimler, Deutsche Telekom and Cofiroute (a French motorway concessionaire). Big technical difficulties emerged before starting the system operation such that it was shifted to January 2005. The German Federal Government brought the case forward to a Court of Arbitration and the case ended in May 2018 with an agreement of the Consortium to pay 3.2bn EUR to compensate for the lost revenues. The Toll collect company has been taken over temporarily by the Federal government and was planned to be tendered to a private operator later.

Fig. 7.2 Toll collect on-board unit for a 5-axle truck in Germany. *Source* Toll Collect (2019a)

(a) *Technical system*

The payment system leaves two possibilities of payments: First a manual payment at borders or gasoline stations or internet booking, and secondly an electronic payment using GPS satellite communication (see Fig. 7.2). An on-board unit has to be installed in the vehicle, which is rented to the haulage company cost-free, while the costs of installation have to be covered by the vehicle owner. The latter will furthermore not receive compensation for the idle time of the vehicle (about half a day). Toll collect sends invoices to vehicle owners with a record of kilometers driven, comparable to the payment systems of telecom companies.

While the kilometers driven on the motorways are registered by satellite technology, control of payment is performed by video cameras taking photos of number plates. If a truck is suspected of non-payment or of manipulation of the vehicle's on-board unit it can be controlled by officers of the Federal Agency for Road Freight Transport (not by the police which controls traffic regulation). While the Swiss and Austrian payment systems show comparatively low costs the German dual system is very costly, with about 17% of revenues in the start-up phase which dropped below 10% (estimated).

(b) *Economic/environmental foundation of charges*

According to EU Directive 1999/62/EG the average allocated costs of HGV on the motorway network are the baseline for setting tolls. This could be varied until 2006 according to environmental characteristics of vehicles (EURO categories) by a maximum of 50% and later by a maximum of 100%. Furthermore, variations according to congestion and for mountainous areas were possible. The German infrastructure cost calculation scheme came out with two major advancements:

First, a full cost calculation was carried out on the basis of the opportunity cost principle. This means that the capital costs were calculated by using optimal replacement and maintenance modelling instead of using past expenditures. This led to much higher results compared with the past expenditures approach (which was originally favored by the EU Commission) because in Germany the replacement and maintenance needs had been neglected for decades. The opportunity costing principle furthermore requires the consideration of interest on capital employed, which for instance implies that (interest) costs of used land are considered, although land as such is not physically consumed by the use of roads.

Secondly, the allocation of costs to the different user categories followed a strict cost responsibility principle. This means that the user pays not only for the costs caused by the traffic activity (marginal costs) but also for the costs which he or she causes by requesting a particular type of capacity supply. Trucks require a different thickness of road layers, width of lanes, stronger bridge construction and wider tunnels. When they are operating on the roads they require more road space than cars. When they are parked at parking lots by the motorways they need more space and land-take. Passenger cars require a different design of road alignments to allow for higher speeds (designs of curvature or exit/entry/intersection construction). The cost allocation scheme applied has been derived from cooperative game theory which explains how different players can be motivated to get together to build and finance a common facility (e.g.: a golf course) and distribute the costs among the players according to the principles of efficiency and fairness.

Figure 7.3 shows the principles of cost allocation.[3] The upper row shows the types of cost as they are defined in the cost accounts: capital costs of earthworks, base courses, binder courses and surface layers, of tunnels and bridges and of equipment

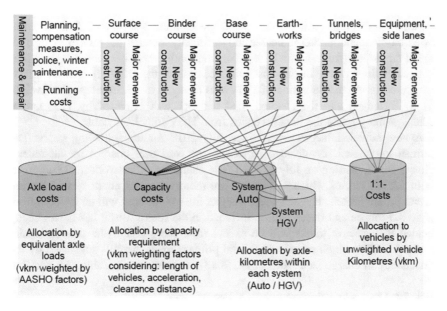

Fig. 7.3 Cost allocation scheme for German motorways. *Source* Rommerskirchen et al. (2009). © The Author(s)

---

[3]For details see Rommerskirchen et al. (2009).

and side lanes, as well as running costs for operation. The cylinders below show the cost drivers, as they are high axle loads, capacity (space) requirements, car specific (high speeds), truck specific (low gradients, lane, tunnel dimensions) and finally common costs which cannot be allocated on the basis of the responsibility principle. Altogether 105 functions were defined and tested to allocate the costs first to HGV and other user categories and secondly to the HGV categories. The HGV categories were defined by the number of axles (three axles, four and more axles).

The charging system has been modified several times since 2005. A major change came after 2011 when the revised Directive 2011/76/EU allowed for adding environmental costs for air pollution and noise. The German charging scheme made use of the option to add air pollution costs while the EU regulations for adding noise costs were so restrictive that they could not be fulfilled by the payment system, i.e. they required a differentiation of noise costs on the basis of exposed population and proven health impacts. The cap values for air pollution costs set by the Directive were so low that little impact followed from their integration into the scheme (between 0 and 3 EU cts per HGV km). In the recent revision of the charging scheme the air pollution charges were increased substantially and a constant value for noise was introduced (see Table 7.4).

This impact could reverse through a change of valuation parameters used for the infrastructure costing. As the interest rates had fallen on the capital market following the economic crisis and the flooding of financial markets with money powered by the European Central Bank, the German Ministry of Transport followed the financial market indicator and drastically lowered the cost of interest on capital employed (from 4.5 to 1.7%). Together with other changes of the previous cost calculations this led to a substantial decrease of the HGV charges for infrastructure costs. As the definition of HGV was changed to >7.5 tonnes gross weight (from 12 tonnes) and the charged road network was extended by 1135 km of federal primary roads the total revenues from HGV tolls kept stable.

A new situation has emerged since July 2018 for the following years. The German Parliament has decided to extend the road user charges for all federal primary roads, starting on July 1, 2018. The charged road network now comprises about 52,000 km, of which 12,900 km are motorways and about 39,000 km are federal primary roads. It is still open whether the weight limit will be lowered from 7.5 to 3.5 tonnes and buses will be included in the future. But it has been already decided that electric trucks will be except from the toll. For the conventionally powered HGV on motorways and federal primary roads the tolls suggested by the consultancy in charge (see Alfen et al., 2018) are exhibited in Tables 7.3 and 7.4.

Table 7.3 Charges for infrastructure costs in Germany since 1/2019

HGV category	Charge for infrastructure costs (cts/vkm)
HGV 7.5 – 12 t	8.0
HGV 12 – 18 t	11.5
HGV 4 axles 18 t	16.0
HGV 4 and more axles, 18 t	17.4

Source The author. See also Toll Collect (2019b)

Table 7.4 Charge for environmental costs in Germany since 1/2019

HGV category	Mark-up for Air pollution (cts/vkm)	Mark-up for noise (cts/vkm)
EURO 0, I	8.5	0.2
EURO II	7.4	0.2
EURO III	6.4	0.2
EURO IV	3.2	0.2
EURO V	2.2	0.2
EURO VI	1.1	0.2

Source The author. See also Toll Collect (2019b)

The mark-ups for air pollution and noise are exhibited separately, because according to EU Directive 2011/76/EU the environmental charge is a separate payment with a different motivation. The tariffs vary from 8.0 to 17.4 cts/km for infrastructure costs and 1.3 to 8.7 cts/km for environmental costs. EURO VI trucks with 40 tonnes of max. gross weight, which predominantly operate on European long-distance routes, will pay 18.7 cts/km. The revenues go to a recently established public Federal Highway Society which will spend the money on operation, maintenance and extension of the federal road network. In contrast to the Austrian ASFINAG, financing investments by loans or credits is not allowed.

7.4 Development of EURO Categories and Interoperable Payment Devices

The underlying definition of EURO categories is given in Table 7.5. In recent years the testing procedures have changed such that the table is highly differentiated and includes a number of contingencies. Interested readers are referred to the internet and publications of the EC Directorate General on the Environment.[4] The EU Commission is presently preparing the next stage of emission standards which will include EURO VII and has started with stakeholder consultations.

With respect to the heterogeneity of technical systems the EU Commission is preparing a solution together with the member states and the industrial suppliers of the payment technologies. The new system, called EETS (European Electronic Tolling Service) will make it possible to apply only one on-board unit for the different electronic tolling systems. This will make operations on international routes through different EU countries much easier and cost efficient (see Fig. 7.4).

---

[4]EU Commission (2019) Emissions in the automotive sector. https://ec.europa.eu/growth/sectors/automotive/environment-protection/emissions_en.

Table 7.5 EURO emission standards for HGV

(a) Steady-state testing for heavy-duty diesel engines

Euro	Date	Test	CO	HC	NO$_x$	PM	PN	Smoke
			(g/kWh)				(1/kWh)	(1/m)
I	1992, ≤ 85 kW	ECE R-49	4.5	1.1	8.0	0.612		
	1992, >85 kW		4.5	1.1	8.0	0.36		
II	1996.10		4.0	1.1	7.0	0.25		
	1998.10		4.0	1.1	7.0	0.15		
III	*1999.10EEV only*	ESC & ELR	*1.5*	*0.25*	*2.0*	*0.02*		*0.15*
IV	2000.10		2.1	0.66	5.0	0.10^a		0.8
V	2005.10		1.5	0.46	3.5	0.02		0.5
VI	2008.10		1.5	0.46	2.0	0.02		0.5
	2013.01	WHSC	1.5	0.13	0.40	0.01	8.0 × 10^{11}	

(b) Transient testing for heavy-duty diesel and gas engines

Euro	Date	Test	CO	NMHC	CH$_4$^{b}	NO$_x$	PM^c	PN^f
			(g/kWh)					(1/kWh)
III	*1999.10 EEV only*	ETC	*3.0*	*0.40*	*0.65*	*2.0*	*0.02*	
	2000.10		5.45	0.78	1.6	5.0	0.16^d	
IV	2005.10		4.0	0.55	1.1	3.5	0.03	
V	2008.10		4.0	0.55	1.1	2.0	0.03	
VI	2013.01	WHTC	4.0	0.16^e	0.5	0.46	0.01	6.0 × 10^{11}

Source The author. See also EU Commission (footnote 5); DieselNet (2019)

^{a}PM = 0.13 g/kWh for engines <0.75 dm^3 swept volume per cylinder and a rated power speed >3000 min^{-1}

^{b}For gas engines only (Euro III–V: NG only; Euro VI: NG + LPG)

^{c}Not applicable for gas fueled engines at the Euro III–IV stages

^{d}PM = 0.21 g/kWh for engines <0.75 dm^3 swept volume per cylinder and a rated power speed >3000 min^{-1}

^{e}THC for diesel engines

^{f}For diesel engines; PN limit for positive ignition engines TBD

7.5 Conclusions Regarding the Effectiveness of Tolling Systems

As it is difficult to evaluate the specific impacts of HGV tolls I take the development of specific charges (cts/km) and total revenues from tolls (bn. EUR per year) as proxy indicators for measuring the additional costs of road hauliers, which influence the road transport tariffs and the decisions of shippers and freight forwarders with respect to their choices of transportation modes. Comparing the developments of the HGV tolls in Switzerland, Austria and Germany and taking the EURO V standard as well as the highest axle or weight category for comparison then the German charges per HGV km remain almost stable (decrease of infrastructure costs,

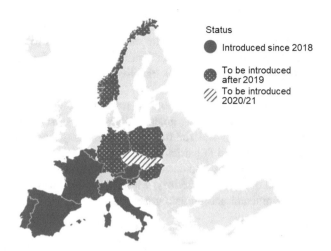

Fig. 7.4 European Electronic Toll Service (EETS). *Source* European Commission (2019)

mark-ups for air pollution), while the Austrian charges have increased modestly, and the Swiss charges have increased substantially.

But it has to be considered that the toll network in Germany has increased by including sections of the federal primary network (1135 km in 2012) in addition to the motorways. Since July 2018, the whole network of federal primaries is included in the toll system, such that the total payments of road hauliers have increased modestly in recent years and will increase drastically beginning with the second half of 2018. It is estimated that the revenues from HGV tolls in Germany will increase by about 2 bn. EUR per year by extending the charged network. As the total revenues in the year 2017 amounted to 4.7 bn. EUR this makes an increase of 43%. The total revenues in the year 2005 were 2.6 bn. EUR, which underlines that despite the very moderate changes of the specific tolls per HGV km almost a doubling of payments was recorded, which stems from the growth of HGV transport, extension of the toll network, the inclusion of HGVs between 7.5 and 12 tonnes gross weight and the mark-ups for the environmental costs, although the latter appear very low.

Also, the neighboring countries recorded close to a doubling of HGV toll revenues from 2005 to 2017, but the reasons for this increase were higher tariffs (in particular in Switzerland) and increased truck km driven (in Austria). Switzerland is the only country which could stabilize the HGV km driven (2006: 2.16, 2016: 2.24 bn. truck.km). In the same time period truck.km increased in Germany from 371 to 432 bn. truck.km (16.5%) and in Austria from 32 to 41 bn. truck.km (28.5%).[5] This indicates that the Swiss transport policy is very strict, following the goal of shifting the growth of freight transport from road to rail. The European

---

[5]Data taken from *EU transport in figures–Statistical Pocketbook* 2017.

transport policy follows the same goal (see the European Commission's White Paper from 2011), but when it comes to implementation it has to make compromises according to the different interests of (until 2019) 28 member states, and this is reflected in the lower environmental effectiveness of HGV tolls in the EU countries.

In this context it has to be added that tolls on traffic on European roads is not the only instrument to achieve the climate goals set. Revisions of standard settings for emissions (EURO categories) and for fuel efficiency (energy consumption, CO_2 production) are on the way and in particular the forthcoming stricter fuel efficiency standards,[6] paired with new test methods (from lab to road) and approval procedures (EU supervision of national practices) will be promising ways to approximate the ambitious environmental and climate goals set.

References

Alfen Consult, AVISO, BUNG (2018) Calculation of road track costs 2018–2022 for Germany. Final report (in German) [Wegekostengutachten 2018–2022 Endbericht]. On behalf of the German Ministry of Transport and Digital Infrastructure, Weimar

ASFINAG (2019) GO toll rates 2019. The latest data is available at the following URL https://www.asfinag.at/toll/go-box-for-hgv-and-bus/

DieselNet (2019) EU: Heavy-Duty Truck and Bus Engines. Emission Standards. https://www.dieselnet.com/standards/eu/hd.php

European Commission (2011) Roadmap to a Single European Transport Area – Towards a competitive and resource efficient transport system (White paper 2011 on competitive and sustainable transport policy). Brussels. Available at the following URL: https://ec.europa.eu/transport/themes/strategies/2011_white_paper_en

European Commission (2015) Road Infrastructure Charging – Heavy Goods Vehicles. https://ec.europa.eu/transport/modes/road/road_charging/charging_hgv_en

European Commission (2017) EU transport in figures–Statistical pocketbook 2017. Brussels. Available at the following URL: https://ec.europa.eu/transport/media/media-corner/publications_en

European Commission (2019) Intelligent transport systems. https://ec.europa.eu/transport/themes/its/studies/european-electronic-toll-service_en

Rommerskirchen S, Rothengatter W, Greinius A, Leypoldt P, Liedtke, G, Scholz A (2009) *Life Cycle Cost Analysis of Infrastructure Networks: The Case of the German Federal Trunk Roads (Karlsruhe Papers in Economic Policy Research Vol.27),* Nomos, Baden-Baden

Swiss Federal Customs Administration (2017) Performance-Related Heavy Vehicle Charge, HVC Overview 2017 Edition. Bern. Available at the following URL: https://www.ezv.admin.ch/ezv/en/home/suche.html#HVC%20Overview

---

[6]For passenger cars the allowed CO_2 emissions will be reduced to 95 g/km after 2021/22 and the EU Commission plans to reduce this limit by 15% until 2025 and 30% until 2030. It is planned to introduce fuel efficiency standards for HGV.

Toll Collect (2019a) Automatic log-on with the On-Board Unit. https://www.toll-collect.de/en/toll_ collect/fahren/einbuchung/automatisch_einbuchen_mit_der_on_board_unit/einbuchung.html
Toll Collect (2019b) Toll rates. https://www.toll-collect.de/en/toll_collect/bezahlen/maut_tarife/ maut_tarife.html

Werner Rothengatter Professor Emeritus, Karlsruhe Institute of Technology, Doctor of Economics, Research interests: transportation economics, assessment, planning.

Werner Rothengatter graduated from Business Engineering at Universität Karlsruhe in 1969. While as a Researcher and Research Assistant at the Institute for Economic Policy Research 1970–1978, he earned the Ph.D. in 1972 and the Habilitation in 1978, both in Economics. He worked as a Professor for Economic Theory at Universität Kiel (1979), Professor for Economic Theory and Policy at Universität Ulm (1979–1986), as a Visiting Professor at Vanderbilt University, Nashville, Tennessee (1982), as Head of the Transport Division at the German Institute for Economic Research (DIW), Berlin (1986–1989), and received a call to the Universität Münster, Institute for Transportation Science (1989). Since 1990, he worked as Professor at Universität Karlsruhe [now the Karlsruhe Institute of Technology (KIT)], Dean of the Faculty of Economics (2003–2004) and Head of the Institute of Economic Policy Research and its Unit of Transport and Communication.

Since April 2009 he has retired from University obligations. He retired as well from the Scientific Advisory Committee of the German Ministry of Transport, Construction and Urban Development, which he had chaired from 2001 to 2002. In 2010 he was honored with the Francqui-Chair for Logistics at the University of Antwerp and in 2013 with the Jules Dupuit Prize of the WCTRS. He was member of the Advisory Committee of Deutsche Bahn AG and a member of the Reform Commission for Large Projects of the German Ministry of Transport, Construction and Urban Development. He is involved in a number of projects of the European Commission and the European Parliament and works as an author, advisor, rapporteur and reviewer. He was President of the World Conference on Transport Research Society (2001–2007) and is still a member of its Steering Committee. He is a member of the Editorial Advisory Board of *Case Studies on Transport Policy* Journals (published by Elsevier) and Editor of the Springer Series on *Transportation Research, Economics and Policy* (together with David Gillen).

Part III
Ecological Balance and Conflicts in the 21st Century

Chapter 8
Panel Discussion

**Chair: Hisae Nakanishi; Designated speakers: Yang Dongyuan,
Lee Schipper, Itsuo Kodama; Panelists: Yoshinori Ishii,
Hans-Peter Dürr, Yoshinori Yasuda, Minoru Kawada,
Yasunobu Iwasaka and Werner Rothengatter**

Chair (Nakanishi): Now, let us start Part III, the panel discussion. I am Hisae Nakanishi of Nagoya University Graduate School of International Development, and I will be acting as chairperson. Earlier today, we had presentations from six professors. First, I would like to ask three designated speakers to give us their comments and opinions. Then, we will have a question-and-answer session. I would like each of the six keynote speakers to answer the questions from the audience. After a free discussion, Prof. Hayashi, executive committee chairman, will summarize the discussion.

In today's presentations, we heard that Nature and resources are limited. I hope today's discussion will be a key to consider them in the future. Prof. Hayashi, please introduce the three speakers.

Hayashi: I am pleased to introduce the speakers. The first is Prof. Yang Dongyuan. He is Vice President of Tongji University, Shanghai, which is a member of an alliance of universities called the International Academic Consortium that also includes Nagoya University. His specialty is transportation engineering, and he is a director of the World Conference on Transport Research Society. He is also general director of the transportation project of Shanghai Expo 2010.

The next speaker is Prof. Lee Schipper. He is a director of the large Embarq program, which is one of the World Resources Institute's most important. As you will see in the brief summary of speakers' careers, he started as a musician, and was a winner of the jazz festival in Notre Dame in the 1960s. He has had a varied career, and is a world-renowned central figure in the fields of energy, and developing countries' transportation and energy environments.

The third speaker is Prof. Itsuo Kodama. He is at present (2005) Head of the Research Institute of Environmental Medicine, Nagoya University. His specialties

Hisae Nakanishi, Professor, Graduate School of Global Studies, Doshisha University, Japan; Professor Emeritus, Graduate School of International Development, Nagoya University, Japan.

© The Author(s), under exclusive licence to Springer Nature Switzerland AG 2020 95
Y. Hayashi et al. (eds.), *Balancing Nature and Civilization—Alternative Sustainability Perspectives from Philosophy to Practice*, SpringerBriefs in Environment, Security, Development and Peace 32, https://doi.org/10.1007/978-3-030-39059-4_8

are cardiac and circulatory organs, and he is now proceeding to develop a new field that encompasses aging and its care. I would certainly welcome his comments from that angle.

Nakanishi: First I would like to ask our speakers to report. Prof. Yang, would you please start?

8.1   First Designated Speaker's Report

Yang Dongyuan
   Vice President, Tongji University

Listening to the guest speakers' presentations, I found myself drawn to two points. One is that all the world's countries have to cooperate with each other regarding the relationship between the environment and humanity, and sustainable development. The other is that we have to build cooperative relationships between various elements of society, considering the viewpoint of sustainability, and not limited to environmental issues.

In addition, what I would like to talk about is my awareness of the issues from a slightly different standpoint. Today's presentations were mainly made from the standpoint of advanced countries. Simply put, sustainable development can be achieved not only through technological progress, but also through progress by the social system as a whole.

The income differential between advanced countries and developing countries is almost the same as to the differential between the eastern and western parts of China. It is about 7 times. Unless we are able to solve this problem, a significant source of social unrest will remain in the future. In the eastern part of China which is developing rapidly, I think it is important to promote low consumption of energy, low resource usage and environment protection based on the idea of sustainable development. On the other hand, in the western part, we have to proceed with provision of infrastructure, education and management that are required in modern society. When we consider the income differential between the eastern and western parts, is it a good policy to keep the current environmental balance in the poor western part where the balance is almost at a primitive stage?

One good example is a very beautiful spot called Jiuzhaigou. Since the spot is blessed with beautiful nature and scenery, many wealthy people come and enjoy this tourist paradise. Local people obtain a small income from the tourists, and try to understand the world through them. However, we cannot assert that this state of affairs constitutes sustainable development in the true sense of the word.

China is now facing several problems. One of them is the energy problem. We estimate that petroleum resources amounting to 700 million tons will be required based on current energy usage in China, which will greatly affect the world. Another big problem is the contradiction between the necessity of increased food production and urbanization. A lot of land area is required to produce food, but an urbanization problem will emerge at the same time. In the future not only China but the entire world will face this big problem.

As for the environmental problem, for example, Hangzhou, which is located in the Zhujiang Delta, is now developing economically, mainly through industry, but this development has brought heavy metal contamination and a chronic shortage of water. China will certainly need the capability to cope with such situations.

As countermeasures for these problems, various means and approaches are needed in the fields of education, regulation and others. However, when we consider sustainable development, the most important thing is people's way of thinking. In the eastern part of China, various ideas are coming up at the moment related to the Shanghai Expo in 2010, and the direction is similar to those in the advanced countries. On the other hand, in the western part, economic development tends to be more emphasized.

There are quite a few problems that China has to solve as a responsible country, and it is highly likely that China will have a major effect on the world. International cooperation is essential to reach a solution. The nurturing of human resources who can properly understand and absorb the past experiences of various countries, and solve various problems is really desirable. There are already some people who have been tackling the problems together with many researchers and engineers in other countries. I would appreciate your kind understanding and cooperation regarding the situation that China is in.

Nakanishi: Thank you very much, Prof. Yang. Using Chinese cases, your report was quite inspirational in the sense of how we should consider the future of our human life and the environment.

8.2 Second Designated Speaker's Report

Lee Schipper
  Project Director, EMBARQ, World Resources Institute for Sustainable Transport

In today's presentations, Prof. Yasuda, Prof. Ishii and other presenters pointed out several problems such as social collapse and energy collapse. There was a story that compared the use of resources to bank robbery, safebreaking. Various kinds of collapse have been revealed in our society. We also received a suggestion from the audience that the market economy might give us some effective feedback in solving these global problems.

So far, energy problems, security in particular, air pollution and global warming have been discussed. We will probably hear later about various problems from the medical viewpoint. Transportation problems were also pointed out. There are various social arguments surrounding the technicalities of adopting advanced fuels, or the sharing of transportation facilities. The pros and cons of these arguments will be discussed in the future.

Various kinds of people live on this congested earth. For example, India is highly densely populated and air pollution has become very serious there. Their situation reminds us of Japan 35 years ago. Japan has solved environmental problems by introducing various regulations. But in many other countries the population density is so high that the progress of air pollution has not yet been stopped.

I think there are still a number of problems with taxation measures. It may be good to tax cigarettes. I think that there should be a principle of taxing something bad, and not taxing something good. In this sense, Singapore is a good example, and the United States is a bad example. Singaporean leaders are well aware that their land is limited, and they have introduced valuable systems, albeit very complex. As a result, Singapore has succeeded in controlling traffic congestion. Not all their systems can be applied to other countries as they are: they have merits and demerits, but generally speaking, Singapore has achieved great success by introducing effective systems such as taxation.

In London, there is a congestion charge system whereby every car entering a designated area, including on high speed roads, must pay a charge. There are many arguments about whether the charge should be fixed or variable. Stockholm now has various policies for levying tax, but I believe it must have been quite difficult for them to reach political agreement for the introduction of such systems.

As for the United States, I think the situation is very bad. One positive example that has shown great progress is in the state of California where the enforcement of careful regulation has improved fuel efficiency. However, in the United States, the priority is a cheap supply of energy. Based on this, huge subsidies are given to public transportation facilities. At present, they are trying to introduce energy-related laws, but in my opinion, these are bad, regressive laws. For example, a large subsidy is offered to ethanol. This ethanol is, of course, not for drinking, but for automobile fuel. In terms of energy ratio, 70% of ethanol fuel is made of fossil fuel. This kind of subsidy system will not increase the use of public transportation or induce the changeover to cars which have a lower environmental load. This are the kinds of subsidies offered in the United States, the world's largest market economy.

China has a great opportunity, I think. Like the United States, China's fuel tax is low, and they have various problems such as the differences in car-ownership tax between Shanghai and other areas. But before road transportation activity reaches a huge scale, there is a possibility of improvement in the future.

For example, highways in cities and elevated highways can be charged, or the tax rate paid for fuel consumption for transportation can be increased.

By learning lessons from what Prof. Ishii, Prof. Yasuda and other presenters have said, China should be able to act wisely and make various improvements. Learning various ways that other countries have tried and adopting them unthinkingly is not enough. Rather, it is important for China to improve them. Otherwise, we are on the road to collapse. It may not be possible to stop the collapse, but we should at least do what we can to resist it.

Nakanishi: Thank you very much, Prof. Schipper.

8.3 Third Designated Speaker's Report

Itsuo Kodama
Head of the Research Institute of Environmental Medicine, Nagoya University

I would like to express my opinion from the viewpoint of the environment and human health. Today, various problems have been discussed: problems of energy, food, lifestyle and others. However, there was little focus on health.

A survey of TV programs in Tokyo for a period of 12 days between August 1 and August 12, 2005 revealed that the number of the programs concerning health was 92, and the number of the programs concerning the environment was 9. In other words, Japanese people now show a marked interest in maintaining health, but do not show much interest in the protection of the environment. This tendency is shown much more clearly in some other regions of Japan. Why it is so may be related to the view Japanese people take towards nature. I would like to ask Prof. Kawada later to explain a little more. In any case, every day if you turn on the TV, you can find a health-related program on a channel or two. Many programs come over as rather threatening, and you may have had experience after a health-related program of being served a food for dinner that you have never eaten before. To that extent, Japanese people show a great interest in health, but their interest in environmental protection is low.

Let us examine the relationship between environmental change (destruction) and health problems. When we look back on the recent past, it can be roughly divided into two periods. The first period is the "Age of Air Pollution," which lasted through the 1990s (the 1980s can also be included). In this age, the person responsible for the pollution (i.e. the cause) could be determined, and the area affected was limited. As you well know, it resulted in Minamata disease, Itai-Itai disease and Yokkaichi asthma. Health problems of this type are now dealt with fairly well due to technological improvement and progress in countermeasures.

Since around the year 1990 what has become conspicuous are the health problems caused by artificial chemicals and life style. In the case of artificial chemicals, those responsible for the contamination are scattered, and the health problems are caused in a wide range of areas. It is therefore difficult to prove a causal relationship. As examples of artificial chemicals, chlorofluorocarbon destroys the ozone layer and increases penetration of the atmosphere by ultraviolet radiation, which accounts for the rapid spread of skin cancer in the southern hemisphere; dioxin is produced by burning garbage in incinerators, causing various problems; asbestos from building materials is known to be deadly; and environmental hormones contained in various agricultural chemicals, medicines and food additives are hormone-disrupting.

Another factor is lifestyle. In this case, the perpetrator is also the victim. In other words, one's own lifestyle damages one's health. Overeating, obesity, lack of exercise, fatigue, mental stress, urban overpopulation and so forth lie behind this. Concepts of health problems caused by lifestyle have been fairly well clarified. The most well-known example is probably metabolic syndrome. This is a combination of obesity, diabetes, excessive fat in the blood and high blood pressure, and it doubles the risk of cardiovascular disease. WHO, the United States and Japan issued diagnosis standards for metabolic syndrome in 1999, 2001, and 2005 respectively. The standards for diagnosis in Japan are a waist over 85 cm for a man, over 90 cm for a woman, slightly elevated neutral fat, low HDL cholesterol, slightly raised blood pressure and slightly high blood sugar levels. If you sit on a sofa and eat ice cream, while using a mobile phone in your hand, the result is sure to be a bulging stomach. It is caused by excess food intake, excessive fat and lack of exercise. When a person becomes fat, it means that each fat cell gets larger, as the number of fat cells remains the same. The nature of the substance secreted from fat cells then changes. As a result, insulin, the most essential hormone which is used by cells to extract energy from blood to use, does not function. This insulin resistance causes diabetes, excessive fat in the blood, high blood pressure and finally arteriosclerosis, and the risk of myocardial infarction and a cerebral stroke more than doubles.

In the United States, metabolic syndrome has increased by a factor of two or three, mainly among young people, in the past 20 years. The same upward trend has also been pointed out in Japan. I think various changes to lifestyles such as high calorie fast food, a lack of exercise, and the spread of TV games etc. have contributed to this tendency.

My research concerns sudden cardiac death. You often hear from your acquaintances of somebody who was healthy until yesterday, but suddenly died, or that somebody suddenly had a heart attack. In these cases someone who looks fine and seems to be enjoying everyday life dies suddenly. 60–70% of them die because of heart trouble. Most of these are heart attacks or other results of coronary disease, but the direct cause is disorder of the rhythm of the heart's pulse. It is estimated that 300,000 people die every year in the United States, and 40–70 thousand in Japan. The number has tripled in the past 10 years. The sudden death is mainly caused by lifestyle and work environment. When long working hours, irregular labor and mental stress are combined, death often results. If a person has metabolic syndrome, the death risk will be 2 to 5 times higher.

In connection with the above, WWF (World Wide Fund for Nature) and the Ministry of the Environment have issued an index called the Ecological Footprint, which shows how much biologically productive area each nation uses per person. For example, for the US it is 12.2 ha. This means that their current lifestyle cannot be maintained unless the US is 2.19 times its size. If this footprint were applied for the Earth, 5.6 Earths would be needed. In the case of Japan, the index is 5.94 ha, requiring 7 Japans. At this rate, three Earths would be needed, Footprints for developing countries are small. For example, Bangladesh is 0.6 ha, which would require 0.28 Earths. In conclusion, I would like to persuade you that the lifestyle of advanced countries will lead to a crisis of the Earth's environment, affecting people's health further. Thank you.

Nakanishi: Thank you, Prof. Kodama.

8.4   Questions and Answers

Nakanishi: Now, we are going to start the question-and-answer session. Questions and answers will be based on the keynote speeches of Parts I and II. I would like to summarize the gist of the speeches as I understood them.

The first point is the viewpoint on how to grasp the real situation of nature and the environment. In understanding the real situation, this means not only present day problems but also the problems of intergenerational ethics, including future generations; what sense of values we hold, how we should think about our current critical situation, and the alternatives we might consider.

This theme came up in Prof. Ishii's presentation about post-oil strategy, and also in Prof. Yasuda's presentation about the change from pasture to farming, and in the information about the state of industry.

The second point is how we actually monitor the environment. This is the problem of how to make an index and manage it. This point can be found in the presentation of Prof. Rothengatter, who talked about tolls in the EU. In connection with this, Prof. Dürr mentioned the necessity of using new indices like energy slaves, 'eco-person' or 'ecoson' as alternative concepts.

The third point is that international conflict and cooperation is involved when we tackle the problems that are actually happening. The traffic tolls in the EU that Prof. Rothengatter mentioned are an example raising these kind of tensions. The problem of *kosa* also relates to this, and we find international cooperation between China and Japan in handling it, and in the international discussions including Mongolia and Korea.

Lastly, I felt that the viewpoint of respecting one's native culture was found in everybody's speech. Coming into contact with a different culture raises the problem of how to understand each other's lifestyle or the sense of value that has arisen from one's native culture. The environment is a transnational problem, and it is now really necessary to find alternative solutions to such problems.

Let us begin the Q and A session. We have received many questions from the audience, so I would like to thank you very much. I am afraid that as we do not have enough time to answer all of the questions, I would like to ask each speaker to roughly summarize the questions and answer them. Prof. Ishii, would you please start?

Ishii: Since I was the first speaker, I have received a variety of questions. They are really a diverse selection. Some of them are quite impassioned. Some people are wondering what we can do; others are thinking very hard. I cannot answer all these questions in three minutes. In a case like this, I make it a rule to answer in a certain pattern.

It is a very gloomy story. Personally, I would like what I say to be wrong. I hope my story does not come true. However, I'm afraid I'm confident that it will. I would like to ask you to think about this yourself from first principles: this is my main comment. I also said that we are in an age of localization or decentralization. Since I grew up in Toyama, I think I know to some extent what regions such as

Toyama are like. Toyama is in a much more difficult situation than Tokyo. It will be really necessary from now on to consider agriculture, decentralization and local production for local consumption. I would like you to look at my website. If you search for my full name "Yoshinori Ishii", the website will be shown. Please look at the website, which also lists all the books I have published.

My final comment is that Nagoya, where I am now, is the home of Toyota. Toyota has excellent management policies. The late vice–president, Mr. Taiichi Ono, is more famous in the United States than in Japan. He said, "Do not waste [*Muda wo suruna*]" This word, *muda*, is already used in roman transliteration in Paul Hawken's book *Natural Capitalism,* which is a very bulky book on the environment. What has this word to do with Toyota's policy? Toyota's policy is not to create any waste, but to make what is needed at the right time in the right quantity. This idea of not creating any waste being applied in the society of the future is very highly appreciated in the United States in books related with the environment. Their policy is not to make a large quantity of something and then push it, but to make the necessary quantity at the right time and pool downstream. This is said to be Toyota's basic policy, which is "Just in time". Since I am in Nagoya, I would like to make this comment to those of you who live in the home of Toyota. "Do not waste" is a very good expression, and means the same thing as *mottainai* that I mentioned as Prime Minister Koizumi's word at the beginning of today's presentation. Pease think about it. Thank you.

Nakanishi: Thank you very much. Prof. Dürr, please.

Dürr: I have received questions such as "What does syntropy mean specifically?" and "How do we measure syntropy?" Syntropy is the same as negative entropy. Entropy means probability, which is a word meaning something may happen. Syntropy is the opposite, that is, "minus probability" which means something may not happen. We measure this in terms of order or level of order.

I also received questions about the energy crisis. Regarding "How can we avoid the energy crisis?", I have a slightly different opinion from that of Prof. Schipper. I think we can solve the problem under the current market system. If the system is wrong, it is a different story. If it is wrong, the first of all we have to recognize that. Sustainability has three different levels: ecological, social and individual. Economic elements are mixed in all these levels. Economy is important, and people are important. When we make these systems, we have to think whether an ecologically correct judgment has been made, considering the quantity of resources being consumed. Otherwise, individuals will not be able to grow as people.

We have taken the wrong way so far. It was as if were afflicted with cancer. Our current problem is how to stop the spread of cancer. We cannot continue to grow as we have been doing, but need to think of a different way. I think to hold down the spread of cancer, we have to slow down the speed of our growth and think in terms of specialization and differentiation. Various species live on the Earth as a result of the processes of specialization and differentiation. But we are now at a time when these species that have developed like this need to cooperate. Of course, there is a principle of competition as well as of cooperation. However, competition itself is not a goal, but a means to an end. In the future, it will be necessary for us to cooperate with the various differentiated species.

I agree with the idea that China, in a sense, combines a developed and a developing country in the same country. However, China has to hold down the spread of cancer and combine the differentiated countries. For this purpose, China has to narrow the gap, which is that between a rich developed country and a poor developing country. And the people in general have to raise their voices, trying to get rid of income differentials between rich and poor. I don't think that is simple. However, with a principle of nonviolence it should be possible to build a world based on this kind of cooperation. A large proportion of the world population is women. We need women to play an important role.

Nakanishi: Thank you very much. Prof. Dürr pointed out at the end of his comment that women's voices should be bigger. I understand that Prof. Yasuda also mentioned the same thing in the morning session. Prof. Yasuda, would you please speak now?

Yasuda: The questions to me mostly overlap with the questions to Prof. Kawada. I would like to ask Prof. Kawada to answer.

Kawada: The questions received are from various angles, but the main point is how we see the issue of the traditional Japanese view of nature. Since I could not discuss this point earlier due to lack of time, I now feel obliged to talk about it.

Regarding the problem of the Japanese view of nature and sense of ethics, Yanagita's arguments are suggestive. First of all, as I mentioned earlier, we have a sense that nature is guarded or protected by something divine. In other words, we have a vague notion that nature is infinite or limitless. So, from that viewpoint, the idea that nature is limited becomes weak. As mentioned earlier, with regard to environmental pollution itself, responsibility lies in the first instance with the offending enterprise. Admitting that, the reason that the government and the local people could not stop the pollution may be connected to the weak idea about nature. This environmental pollution emerged starting in the middle of the 1960s in parallel with the collapse of the *ujigami* faith.

Next is the sense of ethics. I introduced earlier Yanagita's description of Miyazaki, Kyushu, with regard to intergenerational ethics. It is such a vivid and beautiful description that you will feel you would like to act like this. However, the general route of intergenerational ethics is not limited to the idea of deriving pleasure from trying to give our children and descendants better and fulfilled lives; we also have an intergenerational ethics to "sustain the world." In trying to find a reason for the existence of this world, this is the route to find the meaning for your own life or a morality of life derived from the meaning of the world.

When Yanagita developed his myth analysis of the *ujigami* faith, he made some interesting remarks. By analyzing oral literature such as Japanese folk stories and legends, he tried to restructure the original image of the myth in the *ujigami* faith. But, in the myth image, attention was focused on the prosperity of descendants, while questions about the meaning of the world were weak compared to other cultures. Questions such as "What is this world for and where should it go?" are very strong in Chinese culture and in Christianity. From these appear viewpoints for you to find the meaning of your own life, and to sustain this world. Yanagita points out these viewpoints are weak among the Japanese people.

Yanagita himself worked on the Mandate Council of the League of Nations. He repeatedly stressed how important it was to have the viewpoint of how to sustain and guide the world to solve various problems. In addition to the Japanese sense of ethics, in which Japanese showed more interest in society related to descendants, Yanagita said that humans had a lot of problems that had to be solved jointly, although Japanese often tend to think only of their own country. He thought that the League of Nations had to cope with such problems and it was the organization for such purposes. The first issue that arose was how to prevent the next world war. World War I caused unprecedented destruction. The League of Nations somehow tried to prevent the recurrence of such destruction.

Yanagita thought that the League of Nations had to solve the problems of war, peace, labor, and people whose living environment was being destroyed by Westernization, and he also thought that the Japanese people had to be aware of those problems. I think what he pointed out is very important even now. The reason is that at present, in addition to the problems Yanagita listed, the problems that our international society has to face – such as infectious diseases, the environment, and poverty – have increased greatly. That is why I think what he pointed out has still an important meaning even today.

However, some feel that there is a contradiction in what Yanagita says. Yanagita thinks very highly of the distinctive characters of different cultures. This is related to the problem of the environmental conflict mentioned earlier. Since he attached great importance to the distinctive features of each culture, some pointed out the problem of how entirely different cultures can understand each other. Yanagita said that the study of folklore should not be limited to the folklore of one's own country, but must develop into comparative folklore. In other words, we have to aim at a mutual understanding of each other's different cultures. Yanagita has the view that each culture has its own distinctive features, but that each element is common to different cultures. Therefore, his view is based on a kind of universalism, not on a simple devotion to cultural uniqueness.

He was influenced by the anthropology of the time. An example of an element that he listed as common to each culture is the idea of a child being born without a father. In short, this idea is seen in the Bible, and in Japanese folklore and old tales. All over the world there are legends of a child being born after the arrival of a divine spirit. It is a common element. Malinowski found that such a society existed on the Trobriand Islands. This is just one example, but Yanagita stressed the importance of having a universal viewpoint as well as respecting the distinctiveness of each culture.

Nakanishi: Thank you very much. Now, I would like to ask the two professors who made presentations in Part II. Would you please begin, Prof. Iwasaka?

Iwasaka: The questions I received concern on the one hand the current situation of preventing desertification and the spread of *kosa*, and on the other the various standards of *kosa* adopted in different countries. When we consider the first question, of preventive measures, we need to know that it is actually unavoidable that dry areas will appear somewhere in the world, which may sound strange to say; the flow of air gives rise to the dry areas. In any case, it cannot be avoided that the current circulation of air around the earth causes very dry areas in some parts of the world.

As a matter of fact, the Chinese government has issued a warning that the desert near Beijing is advancing 70 km every year, and has taken some preventive measures such as making erosion-control forest and planting trees. Some Japanese NPOs are participating in these projects, but according to the experts this is a very difficult task. To put it in an extreme way, there are quite a few mistaken plantings of trees. It is not enough to plant trees. In other words, plants struggle for a small amount of underground water in dry areas. When plants start to grow in a certain area, sometimes plants in other green areas wither and die. Plants emit water into the air through evaporation and sometimes more underground water dissipates in evaporation than before. So, this is a very difficult problem.

There is another pretty severe view: too many people living there causes the problem. I think this is a delicate problem which some governments may simply not be able to get into. However, the Chinese government's stance is to consider adopting an emigration policy. Therefore, since dry areas are sure to appear, I have the impression that this is unavoidable.

Now, let us discuss *kosa* standards. As an example, Japan established far more standards, much earlier than Korea or China, for the handling of modern meteorology. In Meiji Period, the *kosa* phenomenon was mostly considered in relation to transportation, military affairs and especially fisheries. The most important point is therefore how far we can see. On the other hand, in Korea, their standard is the density of particles floating in the air. They have a great interest in health, and that is why they have such a view towards *kosa*. Among their researchers, there are quite a few experts in health and medicine. China's stance is that they look at nature in dry regions. The wind blowing is very important for them, and they use wind velocity and visibility as the main standards. Each country has set its own standards of *kosa* depending on its interests. It is not realistic to assert that it is necessary to unify the standards, or use the Japanese method. This is the agreed opinion of the people concerned. Thank you.

Nakanishi: Thank you very much. Prof. Rothengatter, would you make your comment?

Rothengatter: First of all, since I am an economist, I would like to raise an objection to what Prof. Dürr said. That is, there is no validity to sustainability. Of course we have to consider the ecological and social aspects, but in order to stop the growth of a cancer, we need to stop the supply of sustenance to it. And clearly a sound economy is necessary to stop that. Many people live in a state of famine in the world, so it is not the case that the economy is not important. This is the first point I want to make. The economy is important, and it is necessary to match the market mechanisms with the ecological aspect. Market mechanisms are a very powerful method. Unless we make use of our economic policies, the environmental policies will not succeed. The important thing is thus not to denounce the market, but to make use of it.

What is the schema of a good market? Regulation and legislation are necessary. Using them, we introduce incentives to promote behaviors to benefit the environment. As an example I talked about the toll system for the automobile traffic. I also talked about the legal implications. In other words, we can work to conserve the environment through taxation systems. This will be the correct incentive. In addition, what I did not explain earlier in my presentation is the green certification prescribed in the Kyoto Protocol. This is an idea to conserve the environment using market mechanisms.

I received a question from the audience: "What should we do about the product liability of an automobile manufacturer in addition to levying taxes on automobiles?" This is a suggestion to lower the environmental load through product liability. It is not included in current European policies. However, I fully agree with such an idea. I would suggest that we incorporate it into the green certification. It can be traded between manufacturers; for example, if a big automobile manufacturer with high energy consumption has to buy a certification from other manufacturers, it will be a mechanism to control CO_2 emission.

The next question is: "Why is there such a wide difference in the prices of on-board units depending upon the country? Why is it so expensive in Germany?" The answer is simple. The technology used in Austria and Switzerland is double track, which is a simple technology that differentiates entrance and exit by means of a gantry. While in Germany they use a different technology. It is the necessity of using it at so many intersections, requiring 10,000 gantries in all, which makes the cost so high.

I will answer the third question simply. The question is: "What is the future of energy?" In other words, how will we cope with the energy crisis, and what role do companies play? I think small and medium-sized companies can play a much more important role. In the future, they can play a very important role with regard to renewable energy in particular. The reason is that even small and medium-sized companies can create renewable energy.

Take wind power generation for example. We can install it in forests, offshore or onshore. Thus, land-owners are able to have wind power generation facilities. Another example is biomass energy. For example, people can make biomass in farms, and thence bio fuels and bio-diesel. In this way, they can make a profit. These are examples of what small and medium-sized companies can do. Farmers also can use biomass for their farm products, and thus can produce organic farm products. In this way, small and medium-sized companies can play an important role.

Nakanishi: Prof. Dürr, do you have any comment, since Prof. Rothengatter expressed a different opinion from yours?

Dürr: I cannot agree with what he said. Of course, the economy is important. The problem is that the schema of a liberal economy is not working well. I think this kind of system has been making a big mistake. In other words, it is based on the idea that human beings are very egotistical from their early stages, and fundamentally, they do not play fair. However, I do not think this is true. I think humans are cooperative from the beginning. It is not good to acquire something from others or deprive other people of their property.

Why did the economy bring about a succession of centralizations, rather than decentralization? Some companies make a huge profit, amounting to 3.5 billion dollars, but on the other hand, many others go bankrupt one after another. In this way, everybody wants to become richer and richer, and make more and more profit. The system itself may not be so bad, but there are so many greedy companies. I think humans are interested not only in economic values but also in other values. Of course, we have to eat to live, but do we have to find value only in eating? We certainly need to eat to grow. In a centralized economy it may well be easier to obtain food. I think, however, that this is a big mistake for the economy as a whole. I don't think we should aim only to obtain profit as the simulacrum becomes bigger and bigger.

Nakanishi: Finally, I would like to ask Prof. Minoru Matsuo, ex-president of Nagoya University, to comment on the significance of this symposium.

Matsuo: First, I would like to say what I felt about today's symposium, including my appreciation of it. Speaking of our organization, please note this is the Graduate School of Environmental Studies, not the Graduate School of Environmental Sciences. When this organization was established, we had a lot of resistance, opposition and demands from the Ministry of Education, Culture, Sports, Science and Technology and examiners. Many scholars and researchers from various fields have gathered, leaving traditional fields of study, to establish a graduate school of environmental studies for the first time in Japan. As one of the founders I am very glad that this graduate school is working steadily.

Another thing I would like to touch on is AC21, one of the cosponsors. This is the consortium that was started just in time for the beginning of the 21st century and the start of the new millennium by a network of over 20 leading universities of the world. The aim of this consortium was to discuss and understand current and future world issues, and send out messages to the world, in addition to discussing educational and academic issues. Today's symposium theme of whether we can design the future of the environment is a challenging one, and I, as a founder of the consortium, was greatly impressed that the excellent speakers, lecturers and panelists have given us wonderful, inspirational talks. I would like to express my sincere appreciation.

Next, I would like to confirm the idea that I have cherished for many years towards the environment, and declare something like a resolution. Today's speakers have all mentioned that in tacking environmental problems, we have to take responsibility and share the burden with much later generations let alone the next three or four generations.

Secondly, the technologies which have a great influence on environmental problems have already transcended the role of tools, to become part of the environment itself. Whether we like it or not, we now have to live, surrounded by software and hardware. In other words, technologies have become more than just tools; they are environmental. As I just mentioned, the environment should be controlled, so we have to confirm that technology should be controlled.

Lastly, my third point is that paradigms have been greatly changing in the 21st century in all aspects such as philosophy, politics, economics, science and technology. Here, I just want to mention one thing, which was also mentioned by today's speakers. That is: we have to move from a limitless paradigm to a limited paradigm. I am an engineer, and our designs up to now suppose that everything is limitless. Our designs have been made on the assumption that energy, materials, and even the purifying ability of the environment are infinite. Now, however, we have to design assuming these are all limited. This is a very difficult situation. I always receive questions as to what we should do, as Prof. Ishii also mentioned a while ago. My answer to this question is that we should do whatever we possibly can, wherever we can. Thank you very much.

Nakanishi: Well, it is almost time to finish. Thank you very much for being with us for such a long time. I think the issue we face is what sense of values we should make from now: in other words, to what extent we have to change our sense of values. Each speaker pointed out that ultimately we cannot continue our present lifestyle.

As ex-president Matsuo said, we should start from whatever we can do. In this sense, if this symposium is a hint for considering how to deal with the environment and live in the future, then this meeting is a success.

In closing, I would like to express my sincere thanks to the keynote speakers from abroad and from distant parts of Japan, and also the executive committee of the Graduate School of Environmental Studies who worked so hard from planning to the running. Thank you very much for your participation.

8.5 About the Participants

Hisae Nakanishi, Professor, Graduate School of Global Studies, Doshisha University; Professor Emeritus, Graduate School of International Development, Nagoya University, PhD in History, Research interests: international politics, Middle Eastern studies, gender and development.

Professor Nakanishi studied at the Department of History, University of California, Los Angeles, was awarded a PhD, and became a professor at the Graduate School of International Development, Nagoya University in 2001. Since April, 2010, she has been Professor of the Graduate School of Global Studies, Doshisha University until today. She was awarded a prize (gender issue program) by the International Peace Research Association in 2005. She was a member of Japan Commission for UNESCO in 2005–2011.

Her publications include: *Islam and Modernity: Aspects of Contemporary Iran* (in Japanese), Fubaisha, 2002; (co-author) *Introduction to Peace Studies – Future Starts Here* (in Japanese), Yuhikaku Publishing, 2004, many other co-authored books such as *Fundamental Theories of Strategic Studies* (in Japanese), Nikkei Publishing Inc., 2011 and *Economic Sanctions under International Laws*, Springer, 2015.

Yang Dongyuan, Professor Emeritus, Tongji University, China, Doctor of Engineering, Research interests: transportation system planning, traffic information engineering, logistics system planning.

Professor Yang graduated from Tongji University in 1982. While he worked and lectured at Tongji University's Department of Traffic Engineering, he completed a doctoral course for Traffic Engineering. He became a professor at the Department of Traffic Engineering, Tongji University in 1994, and the university's vice president in 1997. He also worked as director of Tongji University's College of Transportation Engineering between 2000 and 2001. Professor Yang is an expert in China's urban traffic policies. His publications include *System that Assists Traffic Scheme Strategies*, Tongji University Press, 1997; *Transportation Planning and Management in Continuous Data Environment*, Tongji University Press, 2014; *Urban traffic analysis technology in big data environment*, Tongji University Press, 2015; *Feel the pulse of urban traffic* via *big data*, Tongji University Press, 2017.

The late *Lee Schipper*, Project Director, EMBARQ (World Resources Institute Center for Sustainable Transport), USA, Ph.D. in astrophysics, Research interests: urban policy in developing countries, traffic engineering.

Dr. Schipper studied music (he won at Notre Dame Jazz Festival in 1967) and physics and graduated from the University of California, Berkeley in 1968. He started to work at the Energy and Resources Group, UC Berkeley in 1974. He had worked as Staff Senior Scientist at the Lawrence Berkeley National Laboratory, as a Fulbright scholar in Sweden, as a senior scientist at the International Energy Agency, etc. Since EMBARQ was established within the World Resources Institute in 2002, he had held the position. He was an authority on issues of urban energy and environment in developing countries. He died in 2011.

Itsuo Kodama, born in 1946, Professor Emeritus, Nagoya University; and Senior Scientific Advisor, Suzuken Co., Ltd. He is a Board Member of Japan Heart Foundation, and a Councilor of Suzuken Memorial Foundation. He was a professor at the Research Institute of Environmental Medicine (RIEM), Nagoya University in 1993–2010, and the Director of RIEM in 2004–2008.

His major research interests are electrophysiology of the heart, especially on the mechanism and treatment of cardiac arrhythmias. In collaboration with the University of Tokyo, he developed an optical mapping system of the world highest fidelity, for the experimental analysis of complex cardiac activation during lethal arrhythmias. His publications include: *Guidelines for drug treatment of arrhythmias* (in Japanese) In *Circulation Journal*, 2004; *Cellular mechanisms of sinoatrial activity* In *Cardiac Electrophysiology (4th edition) from Cell to Bedside*, W.B. Saunders, 2004; and many original and review articles in well-established journals such as Circulation, Circulation Research, Nature Medicine, Cardiovascular Research, American Journal of Physiology-Heart and Circulatory Physiology, and Heart Rhythm.

About the Editors

Yoshitsugu Hayashi (Japan), born in 1951, Professor Emeritus, Nagoya University; and Director, International Research Center for Sustainable Development and Global Smart Cities, Chubu University; Distinguished Visiting Professor, Tsinghua University, China. He is a Full Member of the Club of Rome and President of the Japan Chapter, and also has been President of WCTRS (World Conference on Transport Research Society) till May 2019.

His major fields of research are analysis and modelling of transport – land use interactions and the countermeasure policy to overcome negative impacts of urbanization and motorization. The results are published in such books as *Land Use, Transport and The Environment* (Kluwer, 1996), *Urban Transport and the Environment – An International Perspective* (Elsevier, 2004), *Intercity Transport and Climate Change – Strategies for Reducing the Carbon Footprint* (Springer, 2014), the Japanese Edition of *Factor 5* (Akashi-shoten, 2014) originally authored by Ernst Ulrich von Weizsaecker et al., section author of *Come on: Capitalism, Short-termism, Population and the Destruction of the Planet* (Club of Rome Report, Springer, 2018) edited by Weizsaecker and Wijkman.

Applications to practice include his proposition of rail transit oriented urban reform to overcome Bangkok's hyper congestion as the leader of JICA project in mid-90s, which became the trigger to reverse the budget of road vs. rail from 1:99 in 90s to 82:14 in Transport 2020 Plan. He is also now JICA/JST research project leader of "Smart Transport Strategy for THAILAND 4.0".

Y. Hayashi et al. (eds.), *Balancing Nature and Civilization—Alternative Sustainability Perspectives from Philosophy to Practice*, SpringerBriefs in Environment, Security, Development and Peace 32, https://doi.org/10.1007/978-3-030-39059-4

Masafumi Morisugi (Japan) Born in 1970. Professor at Meijo University, Japan. Doctor of Engineering and Master of Economics (Nagoya University), a visiting scholar at Wuppertal Institute for Climate, Environment and Energy, Germany in year of 2017. Various activities around environmental economics and policies.

His current research interests include evaluation several economic damages due to global warming, sand beach erosion, flood increasing, effects on skiing sites and health damages as heat stress and stroke. These topics have been conducted in such research projects as "The Social Implementation Program on Climate Change Adaptation Technology (SI-CAT)", and several Grant-in-Aid for Scientific Research, the Ministry of Education, Culture, Sports, Science and Technology (MEXT), Japan. The results are published in such a book as *Resilience and Urban Disasters: Surviving Cities (New Horizons in Regional Science)* (K. Borsekova and P. Nijkamp Ed., Edward Elgar, 2019).

Sho-ichi Iwamatsu (Japan) Born in 1971. Associate Professor of the Graduate School of Environmental Studies, Nagoya University. Dr. Iwamatsu graduated from the faculty of engineering, Kyushu University in 1994, and received his doctoral degree (Engineering) from the same university in 2000. He started to work as Assistant Professor at the Graduate School of Environmental Studies, Nagoya University in 2002 and has held his current position since 2005.

His research field is synthetic organic chemistry and current research interests include resource utilization and material transformation for sustainable cities and communities. He has authored various articles including *Open-Cage Fullerenes: Synthesis, Structure, and Molecular Encapsulation* (Synlett, Thieme, 2005), *Endohedral Fullerenes with Neutral Atoms and Molecule* In *Strained Hydrocarbons: Beyond the Van't Hoff and Le Bel Hypothesis* (H. Dodziuk Ed., WILEY-VCH, 2009).

About the Authors

Yoshinori Ishii (Japan), President, Mottainai Society (NPO); Professor Emeritus, University of Tokyo, Research interests: global environmental science, energy and environmental theories, remote sensing, engineering in exploration geophysics.

Dr. Ishii graduated from the Department of Physics, Faculty of Science, University of Tokyo in 1955. After working for oil companies, he first assumed an assistant professorship and then a professorship in 1978 at the Faculty of Engineering, University of Tokyo. He joined the National Institute for Environmental Studies, Japan in 1994 (working as 9th president between 1996 and 1998), and worked as a professor at Toyama University of International Studies (2000–2006). He set up an NPO, Mottainai Society, in 2006 and became its president.

His publications include: *Prosperous Petroleum Age Ends – Where Will Humanity Go?* (in Japanese), Engineering Academy of Japan's Environmental Forum, ed., Maruzen, 2004; *The Peak of Petroleum Has Come – "Japan's Plan B" to Avoid Collapse* (in Japanese), Nikkan Kogyo Shimbun, 2007; and *Oil Peak Triggers Food Crisis* (in Japanese), Nikkan Kogyo Shimbun, 2009.

Websites: http://www.mottainaisociety.org/ (*Mottainai* Society)
The recent state of the author can be found at: https://oilpeak.exblog.jp/

Y. Hayashi et al. (eds.), *Balancing Nature and Civilization—Alternative Sustainability Perspectives from Philosophy to Practice*, SpringerBriefs in Environment, Security, Development and Peace 32, https://doi.org/10.1007/978-3-030-39059-4

The late **Hans-Peter Dürr** (Germany), Director Emeritus, Max Planck Institute for Physics; and Professor Emeritus, Ludwig-Maximilians-University Munich, Doctor of Physics, Research interests: nuclear physics.

Born in Stuttgart, Germany in 1929, Professor Dürr graduated from the University of Stuttgart in 1953, and was awarded a doctoral degree from the University of California, Berkeley in 1956. He worked at Max Planck Institute for Astrophysics, and, as a co-researcher of Werner Karl Heisenberg, contributed to creating the theory of quantum mechanics and a unified field theory. In 1969, he was appointed to a professorship at Ludwig-Maximilians-University Munich. He started to work also at Werner Heisenberg Institute in 1971 (later Max Planck Institute for Physics). He was awarded various highly-regarded prizes including the Right Livelihood Award in 1987. He became a full member of the Club of Rome in 1991, and theorized nuclear threats and the sustainability of Earth Gaia holistically to bring them to people's attention. He edited the Potsdam Manifesto 2005 (50 years after the Russell-Einstein Manifesto 1955).

His publications include: (co-editor) *Werner Heisenberg. Collected Works,* 9 volumes, Piper and Springer, 1985–1993; and (co-editor and co-author) *What Is Life?* World Scientific Publishing, 2002. He died in 2014.

Yoshinori Yasuda (Japan), Director, Research Center for Pan-Pacific Civilizations, Ritsumeikan University; Professor Emeritus, International Research Center for Japanese Studies, Doctor of Science, Research interests: Environmental archaeology, geology.

Professor Yasuda was born in Mie Prefecture, Japan, in 1946, and completed a master's degree at Graduate School of Science, Tohoku University in 1972. He taught at the School of Integrated Arts and Sciences, Hiroshima University, and became a professor at the International Research Center for Japanese Studies in 1994.

He also worked as a visiting professor at Humboldt University of Berlin in 1996 and as a professor at the Graduate School of Science, Kyoto University in 1997. He is a pioneer in Japan in the new field of environmental archeology, and was awarded Chunichi Bunka-sho prize (by Japanese newspaper company Chunichi Shimbun) in 1996 and the Medal of Honour with purple ribbon (by Japanese government) in 2007.

His recent publications include: *Introduction to Environmental Archeology – 20,000 years of Japan's Natural and Environmental History* (in Japanese), Yosensha Publishing, 2007; *Civilization of Rice-farmers and Fishermen – From Yangtze Civilization to Yayoi Culture* (in Japanese), Yuzankaku, 2009.

Minoru Kawada (Japan), Professor Emeritus, Graduate School of Environmental Studies, Nagoya University, Doctor of Law, Research interests: History of political thoughts, history of politics and diplomacy.

Born in 1947, Professor Kawada completed a doctoral course at the Graduate School of Law, Nagoya University and started to teach at the University's School of Law. He later lectured at Nihon Fukushi University before taking up a professorship there. He became a professor at the School of Informatics and Sciences, Nagoya University in 1996, and a professor at the university's Graduate School of Environmental Studies in 2001.

His publications include: *Study on the History of Kunio Yanagita's Thoughts* (in Japanese), Miraisha, 1985; *Kunio Yanagita – World of Native Faiths* (in Japanese), Miraisha, 1992; *Kunio Yanagita – His Life and Thoughts* (in Japanese) Yoshikawa Kobunkan, 1997; *Japan as Viewed by Kunio Yanagita – Folkloristics and Design of Society* (in Japanese), Miraisha, 1998; and *Osachi Hamaguchi and Tetsuzan Nagata* (in Japanese), Kodansha, 2009.

Yasunobu Iwasaka (Japan), Professor Emeritus, Graduate School of Environmental Studies, Nagoya University; former Professor, Frontier Science Organization, Kanazawa University, Doctor of Science, Research interests: atmospheric aerosol, atmospheric physics.

Born in Toyama Prefecture in 1941, Professor Iwasaka graduated from the Faculty of Science, University of Tokyo in 1965 and completed a doctoral course at the university's Graduate School of Science in 1971. He became a professor at the university's Solar-Terrestrial Environment Laboratory in 1989, and at the Graduate School of Environmental Studies in 2001. He became a professor at the Institute of Nature and Environmental Technology, Kanazawa University in 2004. He was awarded a prize by the Meteorological Society of Japan in 1990.

His publications include: *Ozone Hole – Earth's Atmospheric Environment Observed from the Antarctic* (in Japanese), Shokabo, 1990; *Introduction to Environmental Sciences 2 – Atmospheric Environmental Science* (in Japanese), Iwanami Shoten, 2003; *Asian Dust – Looking into its Mysteries* (in Japanese), Kinokuniya, 2006; and (co-editor) *Yellow Sand* (in Japanese), Kokonshoin, 2009.

Werner Rothengatter (Germany), Professor Emeritus, Karlsruhe Institute of Technology, Doctor of Economics, Research interests: transportation economics, assessment, planning.

Professor Rothengatter graduated from Business Engineering at Universität Karlsruhe in 1969. He earned the Ph.D. in 1972 and the Habilitation in 1978, both in Economics at the Institute for Economic Policy Research. Since 1990, he worked as Professor at Universität Karlsruhe (now the Karlsruhe Institute of Technology (KIT)). In 2010 he was honored with the Francqui-Chair for Logistics at the University of Antwerp and in 2013 with the Jules Dupuit Prize of the WCTRS. He was a member of the Advisory Committee of Deutsche Bahn AG and a member of the Reform Commission for Large Projects of the German Ministry of Transport, Construction and Urban Development. He is involved in a number of projects of the European Commission and the European Parliament. He was President of the World Conference on Transport Research Society (2001–2007) and is still a member of its Steering Committee. He is a member of the Editorial Advisory Board of *Case Studies on Transport Policy* Journals (Elsevier) and Editor of the Springer Series on *Transportation Research, Economics and Policy* (together with David Gillen).

Hisae Nakanishi (Japan), Professor, Graduate School of Global Studies, Doshisha University; Professor Emeritus, Graduate School of International Development, Nagoya University, Ph.D. in History, Research interests: international politics, Middle Eastern studies, gender and development.

Professor Nakanishi studied at the Department of History, University of California, Los Angeles, was awarded a Ph.D., and became a professor at the Graduate School of International Development, Nagoya University in 2001. Since April, 2010, she has been Professor of the Graduate School of Global Studies, Doshisha University until today. She was awarded a prize (gender issue program) by the International Peace Research Association in 2005. She was a member of Japan Commission for UNESCO in 2005–2011.

Her publications include: *Islam and Modernity: Aspects of Contemporary Iran* (in Japanese), Fubaisha, 2002; (co-author) *Introduction to Peace Studies – Future Starts Here* (in Japanese), Yuhikaku Publishing, 2004, many other co-authored books such as *Fundamental Theories of Strategic Studies* (in Japanese), Nikkei Publishing Inc., 2011 and *Economic Sanctions under International Laws*, Springer, 2015.

Yang Dongyuan (China), Professor Emeritus, Tongji University, China, Doctor of Engineering, Research interests: transportation system planning, traffic information engineering, logistics system planning.

Professor Yang graduated from Tongji University in 1982. While he worked and lectured at Tongji University's Department of Traffic Engineering, he completed a doctoral course for Traffic Engineering. He became a professor at the Department of Traffic Engineering, Tongji University in 1994, and the university's vice president in 1997. He also worked as director of Tongji University's College of Transportation Engineering between 2000 and 2001. Professor Yang is an expert in China's urban traffic policies.

His publications include *System that Assists Traffic Scheme Strategies*, Tongji University Press, 1997; *Transportation Planning and Management in Continuous Data Environment*, Tongji University Press, 2014; *Urban traffic analysis technology in big data environment*, Tongji University Press, 2015; *Feel the pulse of urban traffic via big data*, Tongji University Press, 2017.

The late **Lee Schipper** (USA), Project Director, EMBARQ (World Resources Institute Center for Sustainable Transport), USA, PhD in astrophysics, Research interests: urban policy in developing countries, traffic engineering.

Dr. Schipper studied music (he won at Notre Dame Jazz Festival in 1967) and physics and graduated from the University of California, Berkeley in 1968. He started to work at the Energy and Resources Group, UC Berkeley in 1974. He had worked as Staff Senior Scientist at the Lawrence Berkeley National Laboratory, as a Fulbright scholar in Sweden, as a senior scientist at the International Energy Agency, etc. Since EMBARQ was established within the World Resources Institute in 2002, he had held the position. He was an authority on issues of urban energy and environment in developing countries. He died in 2011.

Itsuo Kodama (Japan), born in 1946, Professor Emeritus, Nagoya University; and Senior Scientific Advisor, Suzuken Co., Ltd. He is a Board Member of Japan Heart Foundation, and a Councilor of Suzuken Memorial Foundation. He was a professor at the Research Institute of Environmental Medicine (RIEM), Nagoya University in 1993–2010, and the Director of RIEM in 2004–2008.

His major research interests are electrophysiology of the heart, especially on the mechanism and treatment of cardiac arrhythmias. In collaboration with the University of Tokyo, he developed an optical mapping system of the world highest fidelity, for the experimental analysis of complex cardiac activation during lethal arrhythmias.

His publications include: *Guidelines for drug treatment of arrhythmias* (in Japanese) In *Circulation Journal*, 2004; *Cellular mechanisms of sinoatrial activity In Cardiac Electrophysiology (4th edition) from Cell to Bedside*, W.B. Saunders, 2004; and many original and review articles in well-established journals such as Circulation, Circulation Research, Nature Medicine, Cardiovascular Research, American Journal of Physiology-Heart and Circulatory Physiology, and Heart Rhythm.

About the Book

This book is an outcome of an international symposium: Sustainability – Can We Design the Future of Human Life and the Environment? which was held as a satellite event of the "Love the Earth"-Expo 2005 (Aichi, Japan). Each chapter is based on the lecture given by the following eminent researchers: Yoshinori Ishii, Hans-Peter Dürr, Yoshinori Yasuda, Minoru Kawada, Yasunobu Iwasaka, Werner Rothengatter, Hisae Nakanishi, Yang Dongyuan, Lee Schipper, Itsuo Kodama, and Yoshitsugu Hayashi.

In the Part I titled "A Sustainable Relationship between Nature and Humans", we discuss what will become of fossil fuels and petroleum, and what kind of indicators should be used to monitor the energy expended by human society. We then discuss environmental impacts caused by different civilizations and values on Nature and ethics, based on the perspective of environmental archaeology and on the discussions by Kunio Yanagita, the father of Japanese folklore study.

The Part II is titled and shows "International Conflict Concerning Environmental Damage and Its Causes". The Asian dust (*Kosa*) is a typical example of trans-boundary conflicts between nations. Another example can be found in the EU's attempt to put in place a common motorway toll system across EU countries having different geographical and economic conditions. Finally, Part III covers the opinions and further debates on sustainable future earth based on the lectures in Parts I and II.

We hope that great insights in this book will come across to readers, and be of help in steering the world towards a sustainable society in harmony with biosystems on earth.

© The Author(s), under exclusive licence to Springer Nature Switzerland AG 2020   119
Y. Hayashi et al. (eds.), *Balancing Nature and Civilization—Alternative Sustainability Perspectives from Philosophy to Practice*, SpringerBriefs in Environment, Security, Development and Peace 32, https://doi.org/10.1007/978-3-030-39059-4

Printed in the United States
By Bookmasters